The Complete
Tent Book

The Complete Tent Book

Andrew Sugar

 Contemporary Books, Inc.
Chicago

Library of Congress Cataloging in Publication Data

Sugar, Andrew.
  The complete tent book.

  Includes index.
  1. Tents.   2. Camping—Outfits, supplies,
etc.   I. Title.
GV191.76.S82   1979   685'.53   78-24567
ISBN 0-8092-7522-8
ISBN 0-8092-7520-1 pbk.

Published by Contemporary Books, Inc.
180 North Michigan Avenue, Chicago, Illinois 60601
Manufactured in the United States of America
Library of Congress Catalog Card Number: 78-24567
International Standard Book Number: 0-8092-7522-8 (cloth)
                                  0-8092-7520-1 (paper)

Published simultaneously in Canada by
Beaverbooks
953 Dillingham Road
Pickering, Ontario L1W 1Z7
Canada

Dedicated to the tentmakers, who devote their lives to designing, testing, experimenting with, and manufacturing the soft shelters that have given so many pleasurable moments to so many campers.

Contents

The Complete
Tent Book

1

A Short History of Tents

A LIKELY STORY

Joe Oook wasn't a typical Cro-Magnon. In fact, ever since childhood he had been considered rather strange by his tribe. Oook was often in trouble with the tribal council and the chief because he kept inventing things that frightened people. And not only did he create such scary items as "fire," "wheels," "wagons," and "spears," but he didn't care that the rest of the tribe constantly ridiculed him for his ideas or that the chief always destroyed his creations, warning him not to meddle in things that Cro-Magnons weren't meant to know. Instead, Oook went on thinking and inventing, ignoring the tribe's jibes, the chief's threats, and the occasional rock aimed at his skull.

Nor was it different on that fateful day when Oook came up with his greatest invention. Oook had been out hunting, despite the fact that the chief had ordered him not to do so. But rather than being concerned over what punishment the chief would mete out, Oook was revelling in the enjoyment of creation. The chief's baleful glare and deep, raspy, foreboding voice never intruded on Oook's joy of creating something the world had never seen before, something the tribe needed desperately—a portable cave. Oook laughed loudly, his guffaws almost drowned by the rattle of the freezing rain pelting everything around him, as he relished the excitement of designing and redesigning his new invention.

It was almost as much fun as the time he figured out how to bring back the kills of the hunting parties faster and with less loss of meat. Ever since the chief had declared that the hunting parties should kill only the animals large enough to feed everyone, the smaller game like antelope and buffalo had been ignored so that the hunters could concentrate on the huge mastodons—chasing them, poking at them with sharpened sticks, throwing rocks at them until the animals either went over a handy cliff or collapsed from exhaustion. But because it took ten to twelve men to drag one carcass back to the series of caves pockmarking the stubby foot-hills where the tribe lived, much of the meat was rubbed away on the rough terrain. And, since the men had to travel slowly, more was lost to the scavengers who fol-lowed the parties back, darting out from the shadows of the forest or swooping down from the air for a mouthful of meat before being driven off. Finally, since it took so long to haul the kill back, on the particularly long hauls in hot weather something would happen to the meat so that it tasted funny. And since the chief had forbidden anyone to use another of Oook's inventions, "fire," sickness often followed full bellies. But Oook secretly cooked the share of meat he received over a small fire he built in the tiny cave he shared with his wife and two children, Debbie and Oook, Jr., and he and his family never became ill. Naturally, this didn't help his reputa-tion with the tribe members, and there was talk that he was an evil spirit.

It was this rumor that was concerning Oook one day when he happened upon a fallen log and rested one foot on it. The log was dead and branchless, rubbed smooth by the weather and forest rot, but still too heavy to be lifted by the tribe's strongest man. Disgusted with the tribe's attitude, Oook angrily pushed at the log and, to his surprise, it started rolling down a small hill. That intrigued Oook, and an idea burst alive in his mind. An invention that could be used to bring back the tribe's meat faster and with less trouble.

After rejecting many different approaches to the prob-lem, Oook finally worked out a scheme. He cut two round pieces off a fresh log and rubbed them smooth with black rocks. (Oook called the black rocks "flints," although the chief had named them "ouchies.") Then he burned a hole in the center of each piece of wood. Oook decided to call this part of his invention "wheels" because the word just seemed to fit. Oook then connected the wheels with a pole. He chose a pole with a diameter slightly smaller than that of the holes so that the wheels could roll freely. Tying another pole to the center of what he called the "axle" (again, he didn't know why the word fit, but it did), Oook piled heavy stones on that connection. Then he tried pulling the load. When it moved easily, Oook knew that he had a means of bringing back fresher meat.

Oook showed his new idea to the tribal council, and they responded in their typical manner. They laughed. They ridiculed him. And they threw stones at Oook and his wife and children. But Oook persisted in trying to sell them on the idea of using his device for hauling back carcasses until the chief stepped in and vetoed the idea. He pointed out that since Oook wasn't a member of the council, that meant that he wasn't smart enough to invent anything useful. Besides, the chief added with a bellowing voice that echoed from the hills, if Oook's "wagon," as he had named it, was truly worthwhile, it would have been thought of before Oook had, and it would have been invented by someone on the council. Finally, the chief explained that his father and his father's father had always brought their kills back to the tribe in the same manner, by dragging them, and it was blasphemous to think of doing it any other way. So the chief destroyed the wagon by having it chopped up in bits and thrown over a cliff, almost throwing Oook after it when he objected a little too strongly. He also warned Oook for the nth time not to meddle in the unknown.

The same thing had happened to Oook's flint-stone spear, but he was sure that the chief wouldn't reject his portable cave. It had limitless possibilities. He didn't have a name for it yet ("portable cave" seemed too long), but he was sure that it would prove invaluable. It would mean that hunting parties could travel longer distances from the caves during the winter, when the animals moved farther away for their food. Wintertime usually meant hungry time, because the hunters often returned empty-handed, forced back by the harsh winds and the white things Oook called "snow" but the chief called "brrrrs." But with Oook's portable cave, they could travel for days or even weeks and bring back plenty of game without the weather bothering them. It meant that——

A sudden crash from the sky forced an involuntary shiver to race through Oook's body, and he peeked out from under the buffalo hide to glance at the sky. It was still ominously black, as it had been for the last seven days, and the rain, along with those little, hard, white balls that the chief called "whack-whacks" but Oook preferred to call "hail," fell off and on. Another crash, this one accompanied by one of those scary white streaks that zigzagged across the sky, and Oook shivered again, withdrawing under his shelter. His brow creased as he finally began to wonder what kind of punishment the chief would think of because Oook had disobeyed him again. But then Oook's optimism returned, and he decided that there probably wouldn't be any punishment. Not after he showed the chief his portable cave. Another thundering boom from above, and Oook trembled, remembering the chief's order.

When the black clouds had first rolled in from the mountains that towered over the caves, the chief, aided by his brother-in-law (who happened to be the tribe's shaman), interpreted the rainy weather as a sign that the Sky God didn't want anyone to go hunting. So he had ordered the entire tribe to remain in their caves until the Sky God brought back the round, red ball that Oook called the "sun" but the chief insisted on calling "oh-oh."

Consequently, like any law-abiding tribal member, Oook had stayed in his tiny cave with his family and gnawed on meatless bones, waiting for the sun (or the "oh-oh") to come back. But, although the rain stopped for hours at a time, the sun never appeared, and the chief didn't rescind his order to wait until it did. The black clouds remained as if stuck to the tops of the bare trees, and the water poured from the sky for hours at a time, drenching everything. Even when it stopped, the ground was too saturated to dry out before the next deluge began. Finally, on the seventh night, Oook could no longer stand the sight and sound of his children and wife crying from hunger. So, picking up one of the extra flint-spears he had hidden in the cave, Oook snuck out of the tribal area to hunt.

It had stopped raining as Oook tramped through the forest, but the wind had gotten colder and it smelled of more rain and perhaps the first snow. Fortunately, the rain held off all night, and in the morning Oook spotted a large buffalo rear-guarding a slow moving herd and killed it with his spear. Fearing the rain would start at any moment, Oook butchered and skinned the buffalo as quickly as he could. He draped the hide over bare branches to dry so that he could use it as a drag-bag to haul back most of his kill to his family. The hide was just about ready when the clouds exploded and the rain poured down heavier than before, this time mixed with hailstones. Frantically, Oook glanced around for some refuge from the freezing rain and hail that clung to the fur wrapped around his body. But there wasn't anything to shield him from the storm.

If it had been spring or summer or even the before-wintertime, he could have sought some shelter under the trees. But now the trees were bare, and there was nothing to hinder the rain and hail from beating on Oook, chilling him to the bone. He considered trudging through the storm with his kill, but it was pouring so hard that he was afraid that he might get lost. Then Oook noticed the buffalo skin still draped over the bare tree limbs. Although the rain had soaked the skin again, the ground underneath the hide was comparatively dry, and it was free from those eye-smarting hailstones. So Oook crawled under the hide and squatted there as he waited for the rain to stop.

As Oook waited, he began to think. Staring up at the skin, he noticed that it leaked in some spots, but it kept out most of the rain and all the hail. A thought began to take shape in Oook's mind: Why couldn't hunters take along dry hides to drape over trees to keep them from getting wet, even in the summer? A blast of cold wind sneaked under the skin, and its icy fingers bit through the fur Oook was wearing. But instead of making him uncomfortable, it triggered even more innovative thoughts. Why not sew several hides together so that there would be four sides to the . . . the . . . Oook needed a name for his idea, but he couldn't think of one. Anyway, if the . . . the . . . whatever had sides, then the top would be a . . . a . . . "roof." This new name sprang to Oook's mind and, like the other names he had made up, it just seemed to fit. But he still couldn't come up with one for his invention. Not yet, anyway. He knew he wou——

Another thought shimmered alive in Oook's brain. There might be times, Oook thought, when there wouldn't be trees with branches at the right height to hang the skins on, so why not take along light poles to make a . . . a . . . "framework"? Oook liked the word and the idea. A wooden framework around which the skins could be fitted and tied. That way, a person could even build a fire inside to keep warm. Oook went on polishing his idea, becoming so absorbed in his invention that he almost didn't notice when the rain finally stopped.

But, remembering his hungry family, Oook threw as much meat as he could into the hide and started for home.

The starving tribe greeted Oook with one of the standard salutations society reserves for innovators: They took away all of his meat, and the chief fined him his catch for disobeying orders. Luckily for Oook, he had figured on something like that and had stashed away enough meat to feed himself and his family before he staggered into the tribal area.

The heavy rains continued for another week, but Oook didn't care. He was too busy perfecting his still nameless invention. First he cut some light, slender poles and burned holes in the tips. He fitted one end of a pole through the hole in another and locked the framework with small wooden pegs he had whittled out of scraps. He discovered that he could make endless varieties of frameworks, but he stuck to one basic design: a "tripod" (his wife came up with that word) structure that proved to be the sturdiest yet lightest of all the designs. Oook's wife stitched together several hides, and then Oook tied them to the frame. When he put a small flap at the top, near the tip of the tripod, Oook found that he could even cook inside the portable cave without being bothered by smoke, because it drifted out of the hole. Finally, Oook asked his wife to sew another flap over the entrance. He could close the flap to keep out the wind if it were cold or open it to let in a breeze if it were warm. Watching his wife working on the flap, Oook got an inspiration. Why not name his portable cave after his wife? Why not call it "Tent Oook"?

He shook his head. It didn't sound right—too long. Then Oook grinned. He would use just her first name.

So Oook called his new invention the "tent."

When Oook showed his invention to the tribal council and the chief, it was immediately rejected. Instead of recognizing the limitless possibilities of the tent, the chief gave Oook two choices: (1) be wrapped up in his stupid tent and tossed off one of the mountains, or (2) take the tent and his family and leave the tribal area for good. Because the snows had already started, the chief thought either choice was a death sentence. But Oook knew that with his new invention he and his family could survive. So he packed up his tent and his family and, wishing he had his wagon or had had the time to perfect another idea of his, a thing he called a "sled," Oook set off.

Some months later, when the warm time had arrived, Oook and his family discovered a friendly tribe living in caves in another hillside. This tribe was led by a different kind of chief, a man named Everett. This chief listened to the story of how Oook and his family had made it across the plains separating the two mountain ranges during the snows by living in the tent. Everett was so impressed with the tent that he made Oook and his family full-fledged members of the tribe and decreed that Oook didn't have to hunt anymore. All he had to do was make tents for the other hunters, and he would receive a full share of the food.

It would be nice to end Oook's story by saying that he went on to become the leader of the tribe when Everett died, as well as the founder and president of the Oook Tent Design and Manufacturing Company, Ltd. But such was not his lot.

Instead, according to drawings found on the sides of caves in southern Europe where Oook lived, he was killed a few months after becoming a member of his new tribe. Oook was experimenting with a new tent design, a long, squatty model that was very light and meant for sleeping only. A pack of wolves mistook Oook's tent for a sleeping buffalo and attacked, eating him before Oook knew what was happening.

Unfortunately, Oook's wife, Tent, wasn't as good a businessperson as Oook, and soon the whole tribe learned how to make their own tents. Therefore, Oook's role as the inventor of the tent was lost to history until recently, when a team of African anthropologists discovered the above-mentioned cave drawings.

THE FACTS

No one really knows where or when the idea of the tent first wiggled alive in a human brain. It is known, however, that animal skins stretched over wooden frames of varying shapes were used as shelters long before permanent homes were built. Even today, as human beings litter the surface of the moon with space debris, an army, an exploratory team, a gaggle of sheep-herding nomads, or a group of scientists probing the past and present can't move very far into the wilds without some sort of tent to shelter their supplies, their equipment, and themselves.

Of course the English word *tent* didn't stem from a Cro-Magnon's first name. It came from the Middle English *tente*, which is derived from the Latin word *tenta*, the feminine form of *tentus*, which means "stretched" or "extended" or "pavilion." The French call a tent a *tente;* the Spanish, *tienda* or *carpa*. The Italian word for tent is *tenda*, and in German it's *Zelt*. *Yurt* is a Turkish word that means "dwelling," and it is used to describe the type of tents used by the Mongols when they were conquering half the world. Virtually every language and every culture has a word that means "tent," and the words, like the structures they name, have migrated from one culture to another evolving along the way.

Tent-like structures have shown up in cave paintings, mostly in the form of lean-tos. One of the first mentions of tents in written form is in the Bible. In Genesis 4:20, Jabal, who lived around 4000 B.C., is referred to as "the father of such as dwell in tents and of such as have cattle." Noah "was within his tent" when warned about the storm, and Lot "had tents," a sign of wealth at the time. Abraham was "at the door of his tent" when he received heavenly messages from angels, and Moses was given instructions on how to make a certain kind of tent to shelter his people during the Exodus: "And thou shalt make a covering for the tent of rams' skins dyed red and a covering above of badgers' skins." Even tentmakers are mentioned in the Bible. In Acts 18:1, 2, and 3, it's written that "Paul came to Corinth; and found a certain Jew named Aquila . . . and because he was of the same craft, he abode with them, and wrought: for by their occupation they were tentmakers."

Ancient references to tents are found in Homer, and Assyrian tent structures are depicted in the sculptured monuments of Nineveh, the ancient capital of Assyria, which had its greatest moments around 615 B.C. Another sculptured tribute to tents is found on the Column of Aurelius (also known as the Column of Antoninus), erected by the Roman Senate about 175 A.D. Much of the detail commemorates the military accomplishments of Marcus Aurelius, but in the background of some of the carvings are examples of tents. Since the Roman military was constantly on the move, these portable structures were an important part of the supplies. As explained in *Roman Antiquities:* "The tents were covered with leather or skins and extended with ropes. In each tent were usually ten soldiers along with the decan, or petty officer, who commanded them." This size of tent was reserved for extremely long spells of bad weather and for unusually long encampments, because the Roman foot soldiers, like all the lower-ranked men of the day, slept outside as long as weather permitted and moved inside tents only when it got too bad for man or beast. The basic framework of the Roman soldier's

Figure 1C

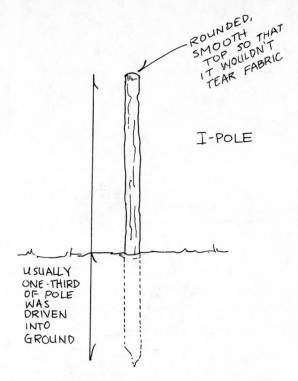

ROUNDED, SMOOTH TOP SO THAT IT WOULDN'T TEAR FABRIC

I-POLE

USUALLY ONE-THIRD OF POLE WAS DRIVEN INTO GROUND

USUALLY 9 POLES WERE USED, ALTHOUGH SOMETIMES 12 POLES WERE UTILIZED...

Figure 1D

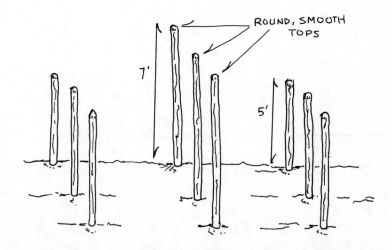

ROUND, SMOOTH TOPS

7'

5'

Figure 1E

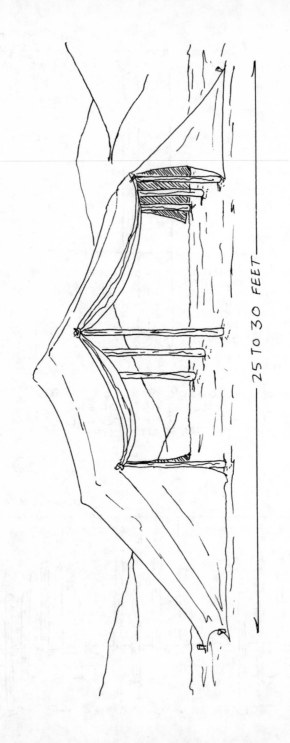

25 TO 30 FEET

tent was the I-pole arrangement (see Figure 1C), simply a single pole driven into the earth. The fabric was draped over it and staked to the ground, forming a pyramidal shape. This primitive frame had been used for centuries when the Romans adopted it, and it is still used today.

The easy-to-erect one-pole framework was also used by the ancient Arabs, usually thought of as tent people, but they also created more sophisticated structures by adding more poles (see Figure 1D). Whatever the number of poles employed, hides or cloths made of coarse woven wool were draped over the poles and the ancient designs were epitomized in the desert tent (see Figure 1E). The Arabs also used *cilicium,* a material made out of goats' hair, and as time went on, they used a more finely woven wool they dyed to add color. But it was the heavy, coarse, rough material, the sackcloth mentioned so often in the Bible, that kept out the sun, the heat, the fine grains of sand, and the wind; so the colorful layers of thinner *cilicium* were used over the heavy wool.

According to *Tents and Tent Life,* a 19th-century text, a very sophisticated tent made its appearance around 300 B.C., when the Persians began using a "circular, wooden lath" for a framework. In addition to skins, leather, and woven wool, they also utilized "pieces of felt"— unwoven wool, hair, or fur matted together by heat, moisture, and great pressure. The Persians hung a colorful rug over the entrance for a door (see Figure 1F).

This type of framework formed the basis of the Mongolian *yurt.* Like the Persian model, the *yurt* had a thin willow frame lattice (see Figure 1G) that worked much like a modern-day protective kiddie gate. Both the Mongols and the Persians began their tent by driving two poles into the ground and connecting them with another short pole to make a rudimentary doorway (see Figure 1H). The lattice framework was then stretched out and one end tied to one driven pole and the other to the

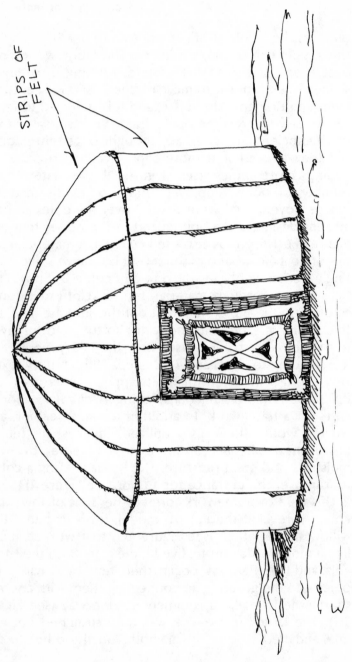

STRIPS OF
FELT

PERSIAN TENT

Figure 1F

second (see Figure 1I). Roof poles were lashed to the top ends of the lattice (see Figure 1J) and the other ends inserted into a metal ring (see Figure 1K). This put the poles under compression, providing a tight structure for the fabric to be tied to (see Figure 1L). The opening caused by the ring also acted as a smoke hole for inside cooking. The tent was about eight feet in diameter and housed several men and their weapons.

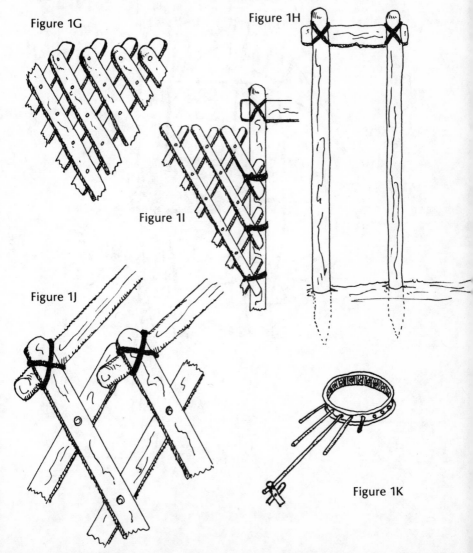

Figure 1G

Figure 1H

Figure 1I

Figure 1J

Figure 1K

Figure 1L

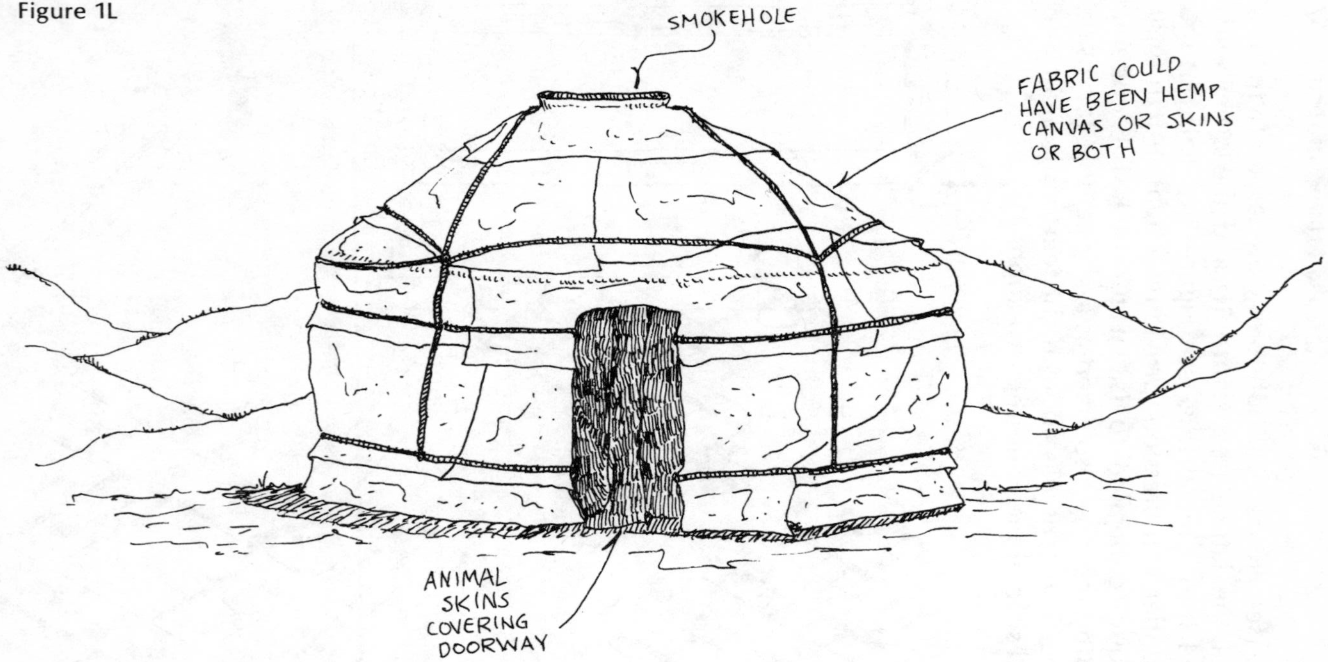

SMOKEHOLE

FABRIC COULD HAVE BEEN HEMP CANVAS OR SKINS OR BOTH

ANIMAL SKINS COVERING DOORWAY

MONGOL TENT

Both the Persian and the Mongolian versions were highly wind resistant, were cool in the daytime and warm at night, and were easily and quickly disassembled and carried on the backs of animals or men. Although some 1,500 years separated the Persian and the Mongolian tents, very few modifications were made in the design. In fact, there are many tents on the market today patterned after the same basic ideas.

The single I-pole construction was improved by various cultures and at various times. The Chinese military of the 15th century was credited with the innovation of connecting two I-poles, used front and back, with a ridge pole (see Figure 1M). This frame allowed for a longer, more stable tent and one that could be quickly set up and torn down, a necessity with the military. The Chinese also improved the fabric—first, by using a hemp canvas, and eventually going to a linen cotton canvas with a very tight weave, making the fabric highly water resistant. The Chinese tent was typically a wedge, A-shape, or inverted "V" tent with the sides staked down, looking much like today's double I-pole tents.

Around the turn of the 18th century some unknown tent designer added another layer of finely woven cotton over the basic tent material to create what is now known as a double-wall construction (see Figure 1N). This kept the people inside the tent even drier, but it wasn't really accepted until the French began using it for officers' tents around 1740. This construction was so successful that eventually most of the governments of the period adopted it as standard issue—for officers. The extra layer of material is now known as a rain fly, and its development was an important chapter in the history of the tent. As pointed out by the *Encyclopedia Methodique:* "The rain easily penetrates the single canvas tents of the soldiers and inferior officers, but the double roof of the marquee (tent with rain fly) effectively keeps out the wet

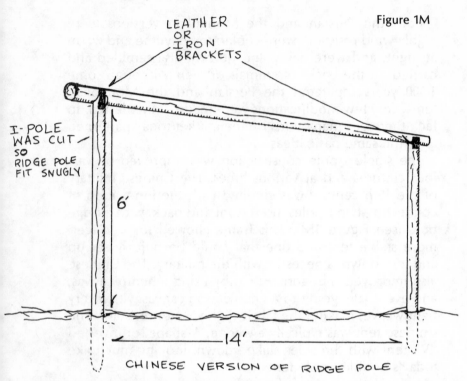

LEATHER
OR
IRON
BRACKETS

Figure 1M

I-POLE
WAS CUT
SO
RIDGE POLE
FIT SNUGLY

6'

14'

CHINESE VERSION OF RIDGE POLE

during all rains, save those of a tropical nature. The double roof renders the marquee much cooler than the other tents which are almost unbearable during the summer months."

To avoid the extra weight of the rain fly, attempts were made to waterproof tents, One such model was issued to British officers in 1840, but it proved to be more uncomfortable than the uncoated fabrics. Because the fabric didn't "breathe," the occupants got just as wet from the condensation of their own exhaled water vapor as the men who didn't have the rubberized cloth. The designers went back to the drawing boards and eventually came up with a waterproofed rain fly to be used with an uncoated fabric for the actual tent. This proved to be the answer, and this same basic idea is used today.

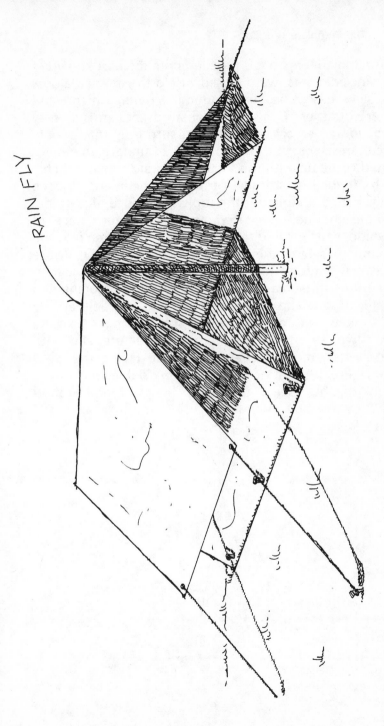

RAIN FLY

FRENCH MARQUEE

Figure 1N

The Indian teepee (also spelled *tipi*), the best known of the ancient tents, wasn't used in North America alone. The same design has been found in Siberia, Lapland, and other countries. This simple yet sturdy design is so easy to build and so basic that it's not surprising that people in cultures separated by thousands of unnavigable miles would come up with the same design at the same time.

The teepee is basically a tripod arrangement of three poles lashed together at the top with a piece of rawhide. Around the tie is secured another, longer, piece of rawhide, which is tied to an "anchor pin" driven into the ground to steady the structure (see Figure 1O). More poles were added to the tripod, but they were leaned against the three main poles, with the last pole being tied to the tripod. The three-legged frame was deliberately built askew, with one of the legs slightly longer than the other two. This gave an angle to the structure, and, with the shorter, more perpendicular side facing the wind (see Figure 1P), the structure was very stable. Skins were sewn together (see Figure 1Q), then wrapped around

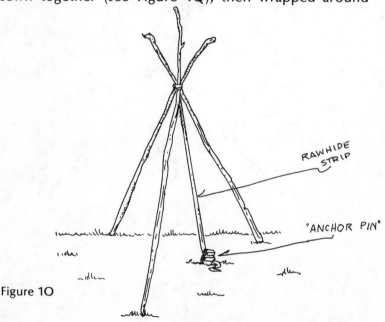

RAWHIDE STRIP

"ANCHOR PIN"

Figure 1O

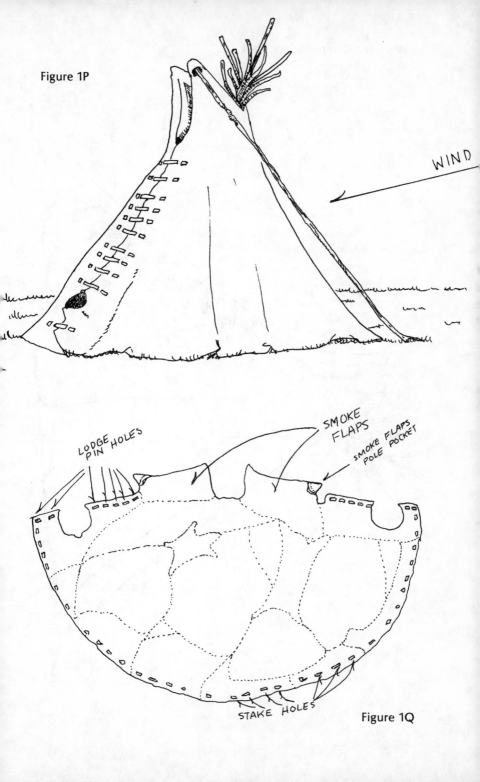

Figure 1P

WIND

LODGE PIN HOLES

SMOKE FLAPS

SMOKE FLAPS POLE POCKET

STAKE HOLES

Figure 1Q

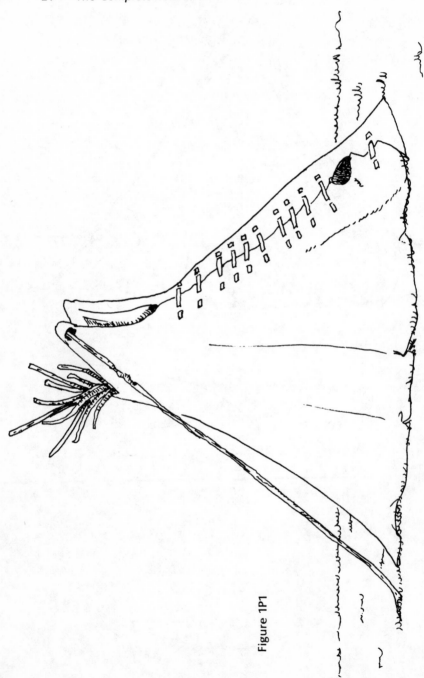

Figure 1P1

and tied to the framework. The fabric then was connected by pins made out of bone. Two outside poles were inserted into the top flaps so that the smokehole could be closed to keep out rain and snow and opened when cooking was being done inside. The finished product looked much the same whether it was in North America or Lapland. Even today, teepees are considered one of the sturdiest, roomiest (the average teepee has an inner diameter of twelve to fifteen feet) and most reliable of all tent designs (see Figure 1P1).

Although the white American never adopted the teepee, instead using the double I-pole arrangement or variations on it, the tent played an important part in American history. The early settlers lived in them for months, or sometimes as long as a year or two, until their permanent houses could be finished. Tents were also important to the armed forces, even when materials were in short supply. According to the *Quartermaster Support of the Army, A History of the Corps, 1775–1939,* published by the Office of the Quartermaster General, tents were in such short supply during the Revolutionary War that the troops used anything to keep out the wind, rain, and snow: "Shelter was as varied among the troops at Boston as uniforms were. The Rev. William Emerson found it diverting to walk among the camps in the Boston area, where every shelter, he thought, depicted the taste of the persons who encamped it.

'Some are made of boards, some of sailcloth, (the Reverend wrote home) and some partly of one and partly of the other. Others are thrown up in a hurry and look as if they could not help it—mere necessity—others are curiously wrought with doors and windows done with wreaths and withes in the manner of a basket. Some are your proper tents and marquees, and look like the regular camp of the enemy. These are the Rhode Islanders, who are furnished with tent equipage from among ourselves and every thing in the most exact English taste.'"

The book goes on to say, "Tents, providing accommodation for six men, were intended to afford shelter for the troops during campaigns though there were seldom enough of them to meet all needs. On the march, wagons carrying baggage, kettles, and tents frequently failed to reach the troops when a halt was called for the night. In such a situation, Sergeant John Smith, retreating through mud and water from King's Ferry to Haverstraw in November 1776, found that he and his comrades were 'obliged to make huts with rails and covered with straw' to sleep under. . . .

"Beginning with the first winter of 1775–76, it became established policy that as soon as the men moved into their barracks or huts, they delivered all tents to the Quartermaster General. Before the next campaign opened, he had them washed, repaired by artisans hired for that purpose, and stored until needed. During the winter months, the Quartermaster General not only directed the salvage of old tents, but, utilizing all the duck that could be obtained, he had new tents manufactured. Replacements were always difficult to obtain and upon occasion that demand could only be satisfied at the expense of other needs. Following the loss of a large amount of tentage during the evacuation of New York City, for example, Congress insisted that 'the soldiers should have Tents if they stripped the Yards of those Continental Frigates and Cruisers that had sails made up.' As a result, committee of Congress reported, 'We have now a parcell of fine vessels laying here useless at a time they might have been most advantageously employed.'"

The tents used at the time were classified in three different categories: common, wall, and hospital. The common tents were the simple inverted "V" designs so popular at the time, but without the rain fly (see the

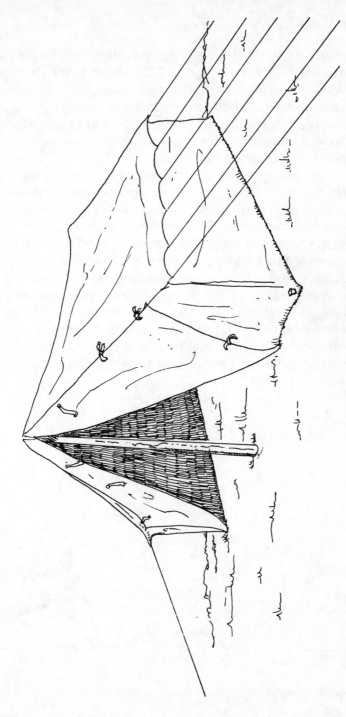

SIDE WALL WITHOUT RAIN FLY

Figure 1R

French model). The wall tent was roomier, larger, and had a rain fly (see Figure 1R). The hospital tent was merely a large circus-tent-type structure that housed many men (the number ranged from twenty-five to one hundred) and was used when quick and easy portability was not a concern. The common and side-wall, or wall, tents were the most common.

One change that occurred during the Mexican war was the method of procurement—from contract-only to a combination of both contracting the tents to various designers around the country and establishing a government-run tentmaking outfit in the Schuylkill Arsenal, near Philadelphia. As reported in the history of the Quartermaster Corps, this method proved superior to contract-only. "By the close of the war, its employees were turning out over 700 common, wall and hospital tents per month. During fiscal year 1848, they made 6,664 common tents, 1,751 wall tents and flies and 192 hospital tents and flies at a lower cost than the Department had

Figure 1S

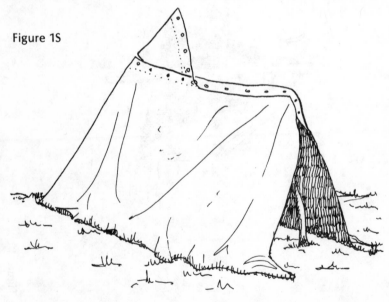

TWO SHELTER HALVES

been able to obtain in its most favorable contract. Only one tentmaker was hired on a monthly basis; the rest—there were 36 employed in 1847—were paid by the piece. Throughout the war, tents continued to be both procured under contract and manufactured at the depot."

The evolution of tentage continued under the guidance of the various governments and under private interests. The French developed a lightweight emergency shelter known as the *d'Abri,* which was a tent made out of two halves that snapped together to form a two-man shelter (see Figure 1S). The Americans developed the Sibley Tent, a center I-pole, circular tent that was large enough to permit ten to fifteen men to sleep inside in a pattern similar to the spokes of a wagon wheel as well as accommodating their gear and a cooking and heating stove (see Figure 1T). Under development since 1855, the tent, named after its inventor, Major H.H. Sibley, was finally incorporated into the Army by 1858, in time for the Civil War, after undergoing rigorous field tests.

As with almost every military item in our past history, the Army was desperately short of all kinds of material when war broke out, and tents were no exception. "No textile was in shorter supply during the Civil War than duck needed for the production of tents," the history on the Quartermaster states. "The supply of tents on hand at the depot in Philadelphia in April, 1861, was soon exhausted. For years, the Army had utilized the common A (inverted "V") or wedge tent, the wall tent, and for winter use, the Sibley tent. These were the only large tents that had stood the test of actual service. There was not enough material in the country, however, to provide tents for all the troops being sent into the field although at the start of the war, some troops were liberally supplied tents. Except for hospital purposes, the use of large tents was not practical in campaigns. To shelter the

Figure 1U. Marines in Peleilu in 1944 used the same design of shelter half as their counterparts in the Civil War. (Courtesy of the National Archives)

Figure 1V. Even before World War II, shelter halves were part of every infantryman's gear as illustrated by this inspection photo taken in the summer of 1941. (Courtesy of the National Archives)

troops on active service, the Quartermaster Department initiated the procurement of tents made on the pattern of the *d'Arbi* tent used by the French Army. The use of the shelter-half (pup tent), so familiar to every American soldier since 1861, was thus introduced in the American Army" (see Figures 1U & 1V).

The shelter-half proved versatile, and, ever since its issue, it has been used in many different ways. In addition to being used as pup tents, they were also used as roofs or walls or even floors in homemade huts or half-bombed-out buildings, a usage that dated back to the beginning of the tent, in 1861.

"When the troops left the training camps for active service in the theater," the *Quartermaster* history said, "they depended for protection in summer on the shelter tent issued to them by the Quartermaster's Department, and after the first few months of the war, in lieu of the large Sibley and wall tents that had been in use, and in winter, on the huts they built when the army went into quarters. Hutting had afforded winter protection ever since the Revolutionary War. The troops usually took up quarters in a well-timbered area, and it was the work of a few days only before a city of log cabins was built. The only difference in the Civil War was that the hut was usually covered with the shelter tent." This same type of usage continued—and still continues—throughout every war the United States has been involved in (see Figure 1W).

After the Civil War, the settling of the West was also aided by the tent. Again, pioneers lived in tents for long periods of time as they built their farm buildings. Tents were important because the settlers built structures for their stock first. Only after their animals were protected did they begin building their own homes, starting from the tent and extending the structure away from it until the canvas shelter was no longer needed.

Figure 1W. The versatility of the shelter half is illustrated here on this Okinawa battlefield as many shelter halves become a roof for a much larger living area whose walls are made up of boxes and ropes.

(Courtesy of the National Archives)

The boom towns that exploded over the American landscape whenever gold, silver, or oil was discovered were composed mostly of tents at first. Occasionally there would be what appeared to be a permanent building—say a storefront or office, made out of scavenged boards and painted—but this was a facade, the main part of the building being a tent (see Figure 1X). Then, as the trains connected these scattered towns and

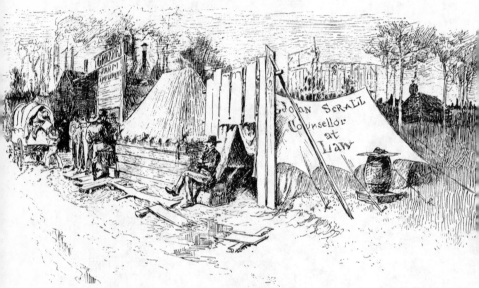

Figure 1X

as lumber became cheaper and more available, permanent homes were built. The tent dwindled in use until it became almost exclusively a military item or a recreational thing reserved for the hearty souls who wanted to experience nature firsthand (see Figures 1Y & 1W).

But the innovations and creative energy expended didn't dwindle, and when the camping boom swept the country in the 1960s and 1970s, the tent remained a highly technical, scientifically designed structure that had generations of experience and creative energy wrapped up along with its poles, fabric, and stakes.

Figure 1Y. Camping has always been a popular American past time as proven by this photograph taken in July, 1938, at Grout Bay Camp in the San Bernardino National Forest in California. (Courtesy of the National Archives)

Unfortunately, there are a lot of inferior, badly designed tents on the market today, insults to the people who have dedicated their lives to designing and redesigning soft shelters. Because of this, many people have had their back-to-nature urges smothered by rotten equipment. But there are ways to determine quality and weed out the junk. This book is one of them. It will show how to find the best buy for your money and how to spot inferior goods. It will also emphasize that the tent has come a long way from the early Joe Oook models and has a long way to go before it becomes obsolete. A very long way to go.

2

Do You Really Need a Tent?

The answer to this question depends on what you plan to do in the outdoors. If you expect your activity to consist of a quick walk down a trail, a trek that takes a half hour or so before you head for the hot dog stand, then you probably won't need a tent. But if you plan to spend all day in the woods, even on established and clearly marked trails in areas with lodges and permanent shelters, a tent or some other kind of protection should be part of your gear. To begin with, you can't depend on permanent shelters or lodges being close at hand when the weather suddenly turns. Also, all the other hikers who haven't thought to carry some shield against inclement weather will be jamming these sanctuaries. Even a lightweight plastic tarp is usually preferable to packed, steamy lodges or three-wall shelters, and they are certainly an improvement over standing in the rain or under a tree. And anyone who neglects to carry shelter because some gas station attendant or coffee shop owner said, "It never rains here this time of year," is asking to be drenched by a downpour.

So even if you're planning to spend just a few hours in the great outdoors, you should have some sort of protection in case nature gets a little nasty. If you are thinking of more than just a few hours away from civilization, if your mind is zeroed in on days or perhaps weeks in the woods, a tent is definitely a must-have item. But buying a tent should not fall into the category of impulse shopping. First you should sit down and compute exactly what you want to spend and then work what you need into the structure of that budget. Before your foot crosses the threshold of any camping supply store, you should decide just what you want out of a tent and how much you're willing to spend to get it.

Perhaps you feel that you really don't know enough about tents to make intelligent choices. If so, don't feel bad. Most people know very little about these soft homes, and that includes the hundreds of thousands of vacation and weekend campers who have plunked down millions of dollars for tents over the past few years without ever really thinking about what they were buying.

For instance, how many individuals ever think of a tent as a highly complicated, highly technical, highly specialized fabric house that is scientifically designed and manufactured? How many think of their tents as the result of years of work, experimentation, innovation, designing and redesigning, and creative energy—culminating in a few poles, some kind of covering, and a particular method for putting the two together? Surveys show that few people ever give much thought to their tents, although they spend days, weeks, or maybe months in them every year. Even many experienced campers don't know all the basic tent terminology and therefore can be intimidated by the technical and semitechnical terms spewed out by tent salesmen. The prospective buyer is inundated with strange-sounding nomenclature, names, and labels and often buys a tent in confusion, without ever understanding what the terms mean. So whether you have been camping for years or are someone who thinks of a tent as an old sheet Mother used to throw over a clothesline when you were eight years old, the first thing you should do is acquaint or reacquaint yourself with all the expressions used in the making and selling of tents. In this way you can protect yourself against overspending because you didn't understand what the catalog meant or what the salesman was saying.

BASIC TENT TERMINOLOGY

Because tents are products of technology and therefore are accompanied by their own technical vocabulary, a simple word may mean one thing to a neophyte but something quite different to a tent designer or intelligent buyer. For instance, *frame* is a simple term that generally means a structure over which something else is placed. But to the serious outdoorsman, *frame* can mean I-pole, A-frame, I-pole/A-frame combination, dome, double A-frame, tunnel, box, umbrella, Baker, Draw-Tite, etc. Each of these frameworks has been designed for a specific function or range of functions. In other words, mention the word *frame* to an experienced, well-informed camper, and you'll be asked, "What kind of frame do you mean?"

The term *fabric* is similarly complex. Tent fabrics include poplin, Army duck, twill, drill, taffeta, or ripstop nylon; the fabric can be waterproofed or fire-resistant; or it can have one or more of a host of other characteristics, all of which huddle together under the generic term *fabric*. As with the term *frame*, each subterm implies a specific purpose, a function, and a comparison of one to another.

As you can see from the terms *frame* and *fabric*, the terminology of tents can be more complicated than it may seem at first glance. This book will explain this terminology in detail in later chapters, but first, to help you fully understand the explanations, some basic tent nomenclature will be explained. These are terms that are used in the manufacturing, designing, selling, and buying of tents, and anyone planning on buying a tent or purchasing accessories for the shelter he or she currently owns should be completely at ease with them. Refer to the accompanying illustration as you read the explanations. It will help you get a concrete idea of each term so that they will soon become as familiar to you as your name.

Ridge Line.   The part of an A-shaped tent that runs from the front support to the rear. The ridge line of most modern tents is cut in a . . .

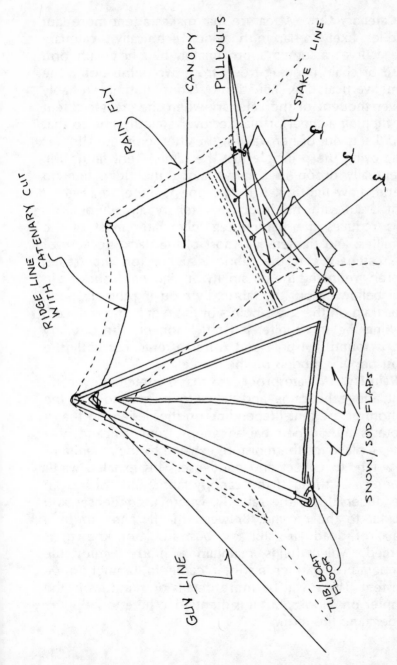

RIDGE LINE
WITH CATENARY CUT

RAIN FLY

CANOPY

PULLOUTS

STAKE LINE

GUY LINE

SNOW/SOD FLAPS

TUB/BOAT
FLOOR

A-FRAME/I-POLE

Figure 2B

Catenary Cut. A feature that makes a tent more taut and less likely to flap in the wind. Technically, a catenary is a "curve" assumed approximately by a heavy uniform cord or chain hanging freely from two points not in the same vertical line. Equation: (Your mother probably never thought of the catenary when she created a tent by flipping a sheet or blanket over a clothesline so that you'd stay out of her way for the afternoon, but she was employing the principle just the same.) Tent fabric will eventually droop in a curve along the ridge line no matter how tightly it is lashed down; therefore, a good tent is cut with the droop, or catenary, already built in. This reduces the natural sag, eliminates most of the wrinkles, and tightens up most of the slack. Because of the catenary cut, the fabric is tighter on the frame, thereby reducing the possibility of high winds sucking in and bellowing out the fabric. By reducing this flapping, the strain on the stress points of the tent is also reduced, making the soft shelter last much longer. The catenary cut is found in I-pole and A-frame tents, currently the most popular models on the market.

Rain Fly. A waterproof tarp that is stretched over the tent, above the frame, and staked to the ground or to the bottom of the tent (depending on the design) so that it protects the air-permeable *canopy* from rain and provides a place for the moisture exhaled by the inhabitants of a tent to collect and condense. This exhaled water vapor eases through the *canopy* fabric and collects on the underside of the rain fly, where it condenses and trickles to the ground. Between the fly and *canopy* a space of dead air (still the best insulator known) is formed, which helps maintain a more comfortable temperature (warmer in winter, cooler in summer) inside the tent. There will be more details on rain flies in the chapter on *Fabrics,* but it is mentioned here to help you understand the term.

Canopy. Simply, the roof of a tent. As will be illustrated in the chapter on *Fabrics,* the canopy should be made of a material that "breathes" and should be used with a rain fly. In fact, both *rain fly* and *canopy* were mentioned so that a constantly used and often misunderstood term could be explained. That term is . . .

Double-Wall Construction. Not quite what it sounds like. Rather than describing two layers of fabric bonded together in some way to form a double wall, this describes the combination of a single-wall tent with a rain fly. Why the employment of a rain fly with a tent is called a double-wall construction isn't really clear. Evidently this is a term that has been around for generations, and the industry has simply adopted it without ever realizing how it confuses a tent buyer, especially a first-timer.

Pullouts. Exactly what the name implies: special pieces of material that are either part of the main tent fabric or are stitched to it. Reinforced grommets in the pullouts allow them to be staked to the ground. This pulls the walls of the tent out, creating a tighter structure and adding a little more room inside.

Ventilation. A very important but little-thought-of feature of a tent. There'll be more about ventilation in the chapter on *Tent Classifications,* but it is mentioned here so that the following terms can be defined . . .

Vents, Tunnel Vents, Exhaust Holes. Quite simply what their names suggest: holes in the fabric to let out stale air and poisonous gases from cookstoves and to allow in fresh air to reduce condensation and discomfort during long stays. The tendency to take vents for granted is one reason for detailing them here as well as in *Tent Classifications.* Vents are important and, since some are designed better than others, they must be considered when buying a tent.

Stake Line. The bottom of a tent, usually one with a *tub,* or *boat, floor,* that has grommets reinforcing the holes through which the legs of the frame pass to the ground. Some stake lines are reinforced themselves with a thick nylon cord that runs around the perimeter of the floor. There is more about how to spot a well or badly designed stake line in the chapter on *Fabrics*.

Tub, or Boat, Floor. All good tents have the floor (usually made of heavy-duty, waterproofed nylon or Army duck) sewn right to the tent. This method of construction is called a tub, or boat, floor. This makes possible a . . .

Self-Standing, or Self-Supporting, Tent. A tent designed so that it doesn't need staking, can be set up just about anywhere, and can be moved without having to be disassembled.

Snow, or Sod, Flaps. Also known as *snow* or *sod frock valances*, these are extra pieces of tent material that can be pulled out and, when secured with snow or sod or rocks, will help steady a tent during gusty winds, when staking is impossible. The flaps also help keep the wind from easing under the floor on the tent, thereby keeping the inside warmer.

Guy Lines. Ropes of heavy-duty nylon (also known as alpine cord), used along with various appliances for tightening the fabric and steadying the tent. Guy lines have more versatility than stakes, as will be explained more fully in *Accessories*.

Figure 2C

DETERMINING YOUR NEED FOR A TENT

An important step in determining your need for a tent is to understand modern attitudes toward camping. Back in the "ancient" times of camping, some twenty-five or so years ago, a camper was likely to be ridiculed by fellow campers for bringing along a tent. He would be called a "sissy" or "weakling" if he didn't entrust his comfort to the luck of the draw in the weather lottery. The argument used to be that if the weather got really bad, the smart camper could make his own shelter out of what grew in the forest. Fortunately, this attitude is no longer a prevalent one. If it were, there would no longer be forests to camp in. Consider: nearly 60,000,000 people stay in our forests and wildlife areas each year; if each one chopped down a couple of trees to build semipermanent shelters, instead of forests we'd have acre after acre of rotting lean-tos protecting nothing but bugs and mushrooms.

Today, happily, no one laughs at a camper who brings along his own shelter even if his stay in the forests is an overnight one. In fact, anyone who makes an impromptu shelter out of the surrounding foliage can be jailed, fined, or both. Or, as often happens, fellow campers may stop him from destroying what it took nature years to grow. No one likes a rapist, whether the victim is a helpless woman or a tree, and the clowns who think it's their right to denude a patch of woods for their comfort are exactly that: rapists and muggers of the forests, criminals who pose a threat to other campers. The true outdoorsman of today is one who brings along his shelter, the individual who can spend a week in the forest and leave without a trace of his stay being apparent to the next camper. So erase the idea of do-it-yourselfing in the woods, unless you're in an emergency situation. The money saved isn't worth it.

If you're an average camper, who settles into the wilderness for a stay of a couple days or a week during the summer months, your tent must satisfy four basic criteria: it should keep you dry if it rains, keep the wind and cold out, keep your body heat in, and provide a barrier to insects. Because all tents basically perform these functions, although some do it better than others, they don't really need to be major concerns when you select *your* tent. Other factors, most of them often overlooked by tent buyers, should be considered when you make your decision.

For instance, a tent that flaps on a windy night can not only lead to deterioration of the tent itself if it's not built to take such abuse, but it can also drive you as crazy as if you were curled up inside a kettle drum during the *1812 Overture*. In other words, your tent may fulfill the four basics but not the *environmental* factor, which is really the fifth basic need.

To take another example: It's common for a person who starts off camping only in the summer in forested areas to get hooked on camping and start taking trips in other seasons and in other terrains. A tent designed only for summer camping in forests won't really fulfill all the needs for a year-round camper. That fifth aspect, environment, pops up again.

So before you look at tents, think of where you may be camping in the future. Get topographical maps of the areas and study the terrain to see if it's hilly and heavily wooded or open and flat. Then match the terrain to the tent. If it's hilly and heavily wooded and you're planning to make trips only in the summer, snow flaps and other wind-resisting accessories won't be necessary. The trees, brush, and hills and valleys will provide natural windbreaks. If the terrain is open or flat or if it is mountainous, a wind-resistant design would be recommended.

Naturally, the weather is also part of the environment,

so you should find out what range of weather conditions you can expect in the area you're planning to visit at the time of year you hope to be there. And remember that even though you may plan to go camping in July or August, you can still be caught in a four-day rain and find yourself confined to a tent that was meant for sleeping only, a shelter in which you can't move around, stand up, or even stretch your muscles. Or you could shiver yourself into a migraine if you neglected to find out that the temperature in that area could suddenly drop twenty or thirty degrees. In the Pacific Northwest, for instance, rains that last for days are not rare, even during the summer months, and a thermometer skidding tens of degrees in an hour or two is also common in the higher elevations. Whether you are heading for the wet Pacific Northwest, the hot and humid Northeast, or the hot and dry Southwest, the environment must be an important consideration in buying a tent, and you should study it thoroughly.

The next factor is tent *size*. If you're a family camper, perhaps you'll want a tent that sleeps five or six. Or you might choose to use several small, two-man tents, one for you and your spouse and the rest for the children. If the children are fairly self-sufficient, they would probably love having their own bedroom under the trees, and this arrangement would give you some privacy as well. Several smaller tents will cost you more than one large one, but the smaller tents have the added benefit of being more portable. If you want togetherness, then you should consider a large model, remembering that you will most likely have to haul it to your destination by car.

The environment is also a factor in determining size. If you're going to be in a rainy area where you may be confined to your tent for several days, you will want one large enough to move around in. Or if you're heading for snow country, splitting up the family may prove a constant inconvenience, as you may have to check the kids a dozen times a night to make sure they're properly protected in their sleeping bags. Here, a large tent is the answer.

Finally, the type of outdoor activity you're planning is one more aspect of computing size. If you are going to spend the entire time in an established campground, complete with running hot and cold water, electricity, and staked-out, level ground, then the large tent may be more suitable. However, if you intend to travel through the backcountry, if you really want to get away from people and civilization, you'll have to carry your home on your back. That means you'll need a small tent designed mostly for sleeping and short durations of protection against inclement weather. Again, you have to match all the factors to your desires.

The last aspect to consider is, naturally, *cost.* To figure this, you should simply match your needs and desires to the functions of various tent designs and then choose among the models that fall within your price range. This may take some wedging and trimming here and there, but if you investigate tents carefully and compare their features with your personal needs, you'll be surprised at how much you can afford.

However, don't be fooled into thinking that more expensive is better. A tent that is designed to be used both in the Arctic Circle and in the Amazon jungle would be nice to own, but why waste the money, especially if it's highly doubtful you'll be traveling in either extreme?

And just as you should be flexible in your desires, cutting out sheer luxury in order to have a tent that is better designed in terms of function, you should also be flexible with your budget. Holding to a set price may mean sacrificing a necessary function to save a few dollars, and those savings may not be worth it. Remember, it wasn't too long ago when you had only two choices in buying a tent: You could purchase either (1) a leaky, bulky canvas model with solid wooden poles that always warped out of shape in the first heavy dew, the type of shelter that had to be hauled in by six adults and an old Ford pickup, or (2) nothing. Once you realize that you have so many options, a few extra dollars may not be so extravagant when you think of the comfort and safety they will buy.

One last hint: When weighing needs and desires against what is available on the market within your budget, ask an experienced camper what kind of tent he uses and why. But beware: don't ask the fashionably chic camper for advice, the person who became drawn to outdoor activity because it was *the* thing to do. These people usually buy very expensive tents for status or because "it looked so cute in the catalog." Since they really don't care about the woods or nature unless some magazine says they should, their opinions wouldn't be worth much. If you don't know an experienced camper, check the local telephone directory. Chances are, there's a backpacking or camping association in your area willing to share information with the beginner.

Note: To obtain topographical maps of where you want to go, first be aware that the U.S. government has the entire nation marked off in sections. To find out which one your intended camping area is in, write the Map Information Office, U.S. Geological Society, Washington, D.C. 20242, and ask for maps of the location you have in mind. You can also write the Washington Distribution Section, U.S. Geological Survey, 1200 South Ends Street, Arlington, Virginia 22202, or the U.S. Geological Survey, Denver, Colorado 80225. Ask for a 1:24,000 map (where one inch on paper equals 24,000 inches on the ground) of the section you wish to visit. Also request any brochures or booklets (there are many) that will explain the symbols used on the maps and will tell you how to permanently mount the maps for field reference. Also be sure to ask for *Silent Guides for Outdoorsmen and Topographic Maps*. It costs only a few cents and is well worth the money.

To determine weather for an area, check with your local bureau of the U.S. Weather Service or write the National Park Service, Washington, D.C. 20240. They will either supply weather reports for any part of the country or tell you how to obtain them. You can also write to the Forest Service, Department of Agriculture, Washington, D.C. 20250, or any of the following regional offices:

Forest Service (Eastern Region)
6816 Market St.
Upper Darby, PA 19082

Forest Service (North Central Region)
710 N. 6th St.
Milwaukee, WI 53203

Forest Service (Southern Region)
50 Seventh St. N.E.
Atlanta, GA 30323

Forest Service (Southwestern Region)
Federal Building
Albuquerque, NM 87101

Forest Service (Intermountain Region)
Forest Service Building
Ogden, UT 84403

Forest Service (Northern Region)
Federal Building
Missoula, MT 59801

Forest Service (Rocky Mountain Region)
Federal Center Building 85
Denver, CO 80225

Forest Service (Pacific Northwest Region)
P.O. Box 3623
Portland, OR 96208

Forest Service (California Region)
630 Sansome Street
San Francisco, CA 94111

You can also write to the Department of Agriculture of the state in which you'll be camping. In addition, dozens of private organizations will furnish camping information for the asking. Many of the following organizations are geared toward backpackers, but whatever your camping style, they can provide you with weather and other environmental information for the areas in which they are located. If the organizations are national in scope, they will also provide information for other parts of the country for a small fee.

Sierra Club
1050 Mills Tower
220 Bush Street
San Francisco, CA 94104

Sierra Trips
P.O. Box 601
Diablo, CA 94528

Appalachian Trail Conference
P.O. Box 236
Harpers Ferry, WV 25425

The Wilderness Society
729 Fifteenth Street, N.W.
Washington, D.C. 20005

American Forest Institute
1619 Massachusetts Ave.
Washington, D.C. 20036

The American Alpine Club
113 East 90th Street
New York, NY 10028

The Colorado Mountain Club
1723 East 16th Avenue
Denver, CO 80218

Pittsburgh Council
American Youth Hostels, Inc.
6300 Fifth Avenue
Pittsburgh, PA 15232

American Youth Hostels, Inc.
20 West 17th Street
New York, NY 10011

Appalachian Mountain Club
5 Joy Street
Boston, MA 02108

The Seattle Mountaineers
P.O. Box 122
Seattle, WA 98111

Michigan Trail Riders Association Inc.
R.D. 2
Box 434
Sutton's Bay, MI 49682

The Green Mountain Club, Inc.
P. O. Box 94
Rutland, VT 05701

Potomac Appalachian Trail Club
1718 N Street, N.W.
Washington, D.C. 20036

Ascutney Trail Association
Windsor, VT 05089

Dartmouth Outing Club
Robinson Hall
Hanover, NH 03755

3

Frames

Once you've decided what your particular needs are, you are ready to look at and compare individual tents to find the one that best suits your needs. The first feature to consider is the bones of the tent, the framework. As already mentioned, the first frames were probably convenient tree limbs over which animal skins were laid to form rudimentary shelters. As the prehistoric outdoorsman became more experienced and more sophisticated in his thinking, he undoubtedly realized that there wouldn't always be tree limbs at the right height to provide adequate shelter, so he bent small saplings to form a tent skeleton. Then it finally dawned on some primitive that there would be times when there wouldn't be trees slender and supple enough to bend, and the first portable frame was devised. Again, readily accessible organic materials were used. Slender wooden poles formed the first lean-tos, which evolved into the more sophisticated teepees.

These first methods of holding up a fabric for shelter didn't change significantly for centuries. Bent saplings became I-poles connected by a ridge pole and the tripod of the teepee became an A-frame, but it's only been in recent years that tent frames have started to become unique constructions and are truly different from what the cavemen used to tie together. The tent buyer of today can select from tent frames designed for heavy winds or no winds; exoskeletons (where the frame is on the outside), permitting more room on the inside; self-standing designs, which need no guying and can literally be picked up and moved without disassembling; and pop-up tents, which spring to their full shape at the release of a trigger.

The materials out of which frames are made have also undergone a great revolution in the last couple of generations. For a long time, frames were made out of just two materials: rope or wood. Today, although rope is still used in some designs, it has changed. Instead of easy-to-rot and quick-to-break cotton or hemp, the rope is now made out of a heavy-duty nylon known as alpine cord. Though it may stretch slightly and requires tightening if the fabric it's supporting isn't moved within a day, alpine cord will hold up enough material to shield ten adults from the elements. Wood has been replaced, for the most part, by lightweight aluminum or fiberglass. Although one can still buy wooden tent poles for a small savings, they aren't worth it. Wooden poles are heavy to begin with and are made even heavier by the various bits of metal (joint endings, tie-on loops, pins, etc.) nailed to them. Though the metal is no longer easy-to-rust steel, it still adds to the weight of the wooden poles. Besides that, anyone who buys wooden tent poles or even wooden tent stakes is adding a lot of weight and giving up a lot of convenience for something that tends to warp, split, crack, swell, and jam inside joints and which has splintered millions of hands in the past.

Instead, the smart camper should go for frames made out of aluminum or fiberglass. They're not only longer lasting, but they are easier to put together. Some companies color code each segment of the pole so that a ten-year-old could put up the framework for a family tent. The same kid could also carry the entire frame if necessary.

Aluminum poles are made out of different alloys and are tempered for additional strength. The alloys are also graded, with, as pointed out by the majority of tent-makers, 6061, 6063, and 7001 being the best grades. Often, the grade number can be found in the company's catalogs, but if it is not given, it will be supplied upon request.

Aluminum poles come in segments for easy storage and transportation, and they are put together in various ways. Some poles merely fit snugly together; some tele-

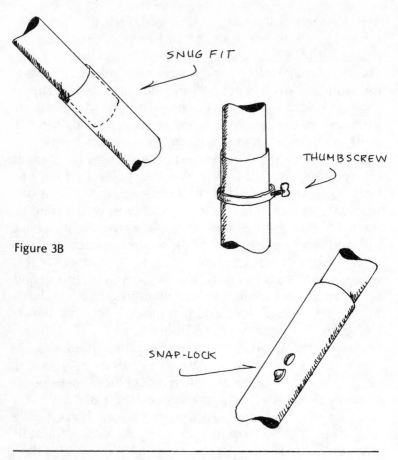

SNUG FIT

THUMBSCREW

Figure 3B

SNAP-LOCK

Figure 3C

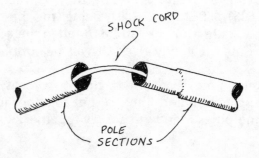

SHOCK CORD

POLE SECTIONS

scope into each other and simply have to be pulled out
to the right height and snap-locked into position (see
Figure 3B). Some come in separate pieces and are snap-
locked together, and some are made to be screwed
together. Also, the better-made poles are sold with
lengths of shock cord running through them (see Figure
3C). Shock cord is usually rubber covered with a heavy-
duty nylon that is woven in such a way as to absorb

Figure 3D

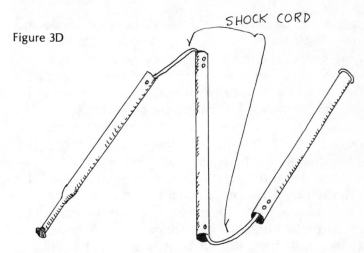

SHOCK CORD

tremendous amounts of strain, such as the sudden
shocks of wind gusts. This cord is threaded through the
various segments of a pole (see Figure 3D) so that the
pieces can't be easily lost and can be assembled faster,
since the sections are always in the proper sequence.

Pole connections can be made out of uncovered aluminum or plastic or of aluminum with a layer of plastic
or brass on top. All four are equally durable and effective. Although bare aluminum joints will freeze together
faster than the other types, this problem can be solved
by proper lubrication (see Chapter 9, *Loving Your Tent*).

Because of the flexibility of fiberglass, solid fiberglass
poles are fast becoming a favorite with both campers and

tent designers. They are light (although slightly heavier than aluminum) and extremely strong, even when bent as far as they can be. Although fiberglass poles are comparatively new on the market, tentmakers have already made improvements. Some are producing hollow fiberglass poles, which are lighter and much more flexible than the solid poles, as well as being as strong as the rigid aluminum. The one drawback is that they are more expensive.

Frames can be divided into two categories: inside and outside. With inside framework, the ones that form the familiar wedge-shaped tents, there's less interior room. However, inside supports are often easier to assemble and easier to tighten, as often becomes necessary if the tent is left standing for long periods of time. The big advantage to exoframes is that there is more interior space and a greater variety of designs are possible. The acute slope of the triangular, or wedge-shaped, tents, which cuts down on standing room, is eliminated by the cylindrical or nearly vertical walls of the exoframes. In dome or tunnel tents, for example, a camper can enjoy up to 40 or 50 percent more room with the same floor area than with the interior suspension systems. And as fiberglass slowly pushes out the more rigid aluminum, look for more exotic designs to begin hitting the market.

Another advantage of the outside framework on some models is that once the tent is erected, the entire structure usually can be picked up and moved if the first location isn't ideal. Also, since the exoframe models are self-standing, they need less guying and staking.

Most of the smaller, backpacker exoframe designs are more wind stable than the more familiar wedge-shaped shelters, and they shed water and snow as easily.

The following is a breakdown of all the types of frames that are being used by today's tent manufacturers, one or two of which should fit right into your plans. Rope will also be mentioned, because it does provide an emergency shelter frame, and it can be used, by those not susceptible to insect attack, to support a lightweight plastic tent. Wood will be ignored, however, because it just isn't worth the savings one *might* gain by substituting wooden frames for aluminum or fiberglass.

Figure 3E

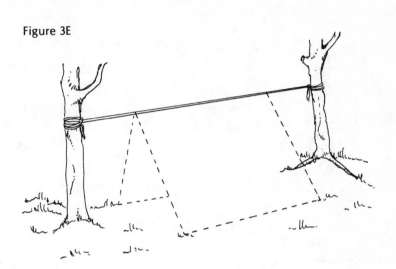

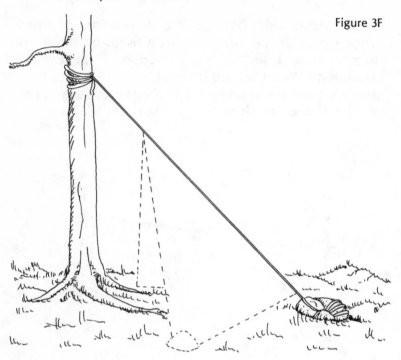

Figure 3F

ROPE

The lightest and least costly of all frames is a simple rope (alpine cord, plastic clothesline, or heavy-duty cotton or hemp) that's long enough to be stretched and tied between two trees, two poles, or even between two backpack frames held upright by rocks (see Figure 3E). Rope can also be used as a frame by tying one end high on a tree limb and anchoring the other end with a rock (see Figure 3F). Whatever way you choose, the idea is to make a semirigid frame over which a tarp, poncho, or piece of plastic can be hung for some protection (see the chapter on *Other Shelters* for complete instructions on the various methods of erecting emergency tents).

Advantages: cheap; lightweight; easy to carry and pack; simple to erect and disassemble.

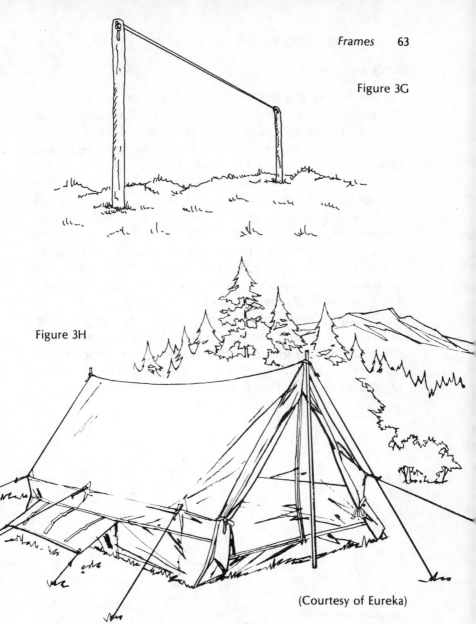

Figure 3G

Figure 3H

(Courtesy of Eureka)

Disadvantages: Any structure erected with rope as the frame offers only minimal protection; even alpine cord can snap or be torn apart in a severe storm; any organic material like cotton or hemp can also mildew or rot.

I-POLE

This usually interior frame is older than rope frames, but poles have been updated so that they are now made out of tempered aluminum. The basic I-pole design consists of one vertical pole either in the middle of the tent or at the entrance. For more stability, two I-poles are used, one in the front and one in the back, connected by a ridge pole or even rope (see Figure 3G). As pointed out in an earlier chapter, one can add as many I-poles as one wishes to make larger and larger structures, but, for the most part, today's tents are usually the double I-pole design with the tent fabric acting as the ridge pole. The double I-pole forms the triangular or wedge shape usually associated with tents (see Figure 3H).

Although the double I-pole frame is surprisingly stable in heavy winds and the triangular shape sheds water and snow, it must be guyed in three different directions with the tent itself pulling in two directions and the guy line pulling in another (see Figure 3I).

Figure 3I

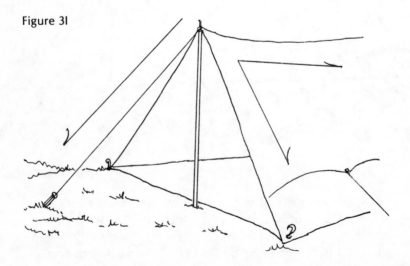

I-pole frames are the skeleton of some of the larger tents on the market, such as the sidewall and the Baker (see Figure 3J). More about these once-condemned but now popular shelters can be found in the chapter on the tent classifications.

Figure 3J

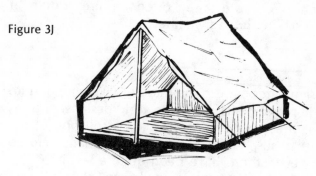

SIDEWALL BY WHITE STAG

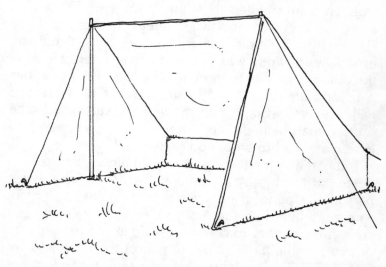

BAKER TENT

Advantages: I-poles are easy to construct and to take down; they are surprisingly wind-resistant, and any damaged poles are relatively easy and cheap to replace.

Disadvantages: Not as wind-stable as some of the

frames described in the following sections; the place-ment of the pole (even if outside the tent) is usually in front of the entrance, making it difficult to get in and out of the shelter; if the I-pole is located in the center, someone will eventually roll or walk into it, causing the whole tent to collapse; they often require additional guying and staking in bad weather.

A-FRAME

This design dates back to the middle 1800s and, despite constant improvements in framework designs, the A-frame, like the more ancient I-pole, is still one of the most popular suspension systems being manufactured today. However, there have been quite a few improve-ments since it was first employed, almost 120 years ago.

For instance, the early A-frame was made out of two wooden planks bolted together at one end, forming an angle of forty to forty-five degrees at the peak (see Figure 3K). The bolt had to be constantly tightened, especially in windy weather, and cracks often started at the hole the bolt went through. The legs of this primitive A-frame were driven into the ground, and the fabric was nailed as well as lashed to the legs. In wet weather the thongs that helped keep the fabric tight tended to loosen and had to be periodically retied.

Today the A-frame is made of aluminum and is con-nected at the top by a permanently hinged joint that is part of the poles, a removable plastic joint, a piece of aluminum bent into the A-shape, or what is known as a cross ridge connection. The cross ridge can be made up of two aluminum rings that are part of the poles and are joined together by alpine cord or it can be a solid piece of aluminum with its ends bent into rings. Either way, the cross ridge connection adds space to the interior of the tent and permits more radical designs without sacrificing

Figure 3K

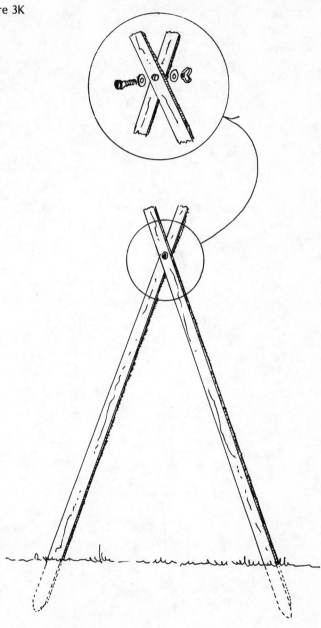

Figure 3L

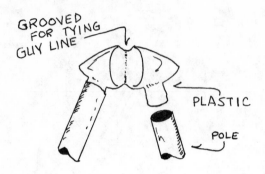

GROOVED FOR TYING GUY LINE

PLASTIC

POLE

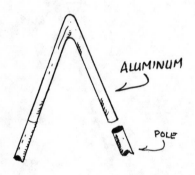

ALUMINUM

POLE

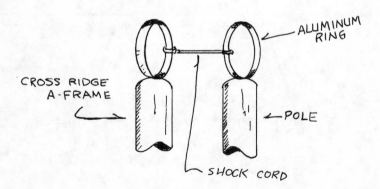

ALUMINUM RING

CROSS RIDGE A-FRAME

POLE

SHOCK CORD

Figure 3M

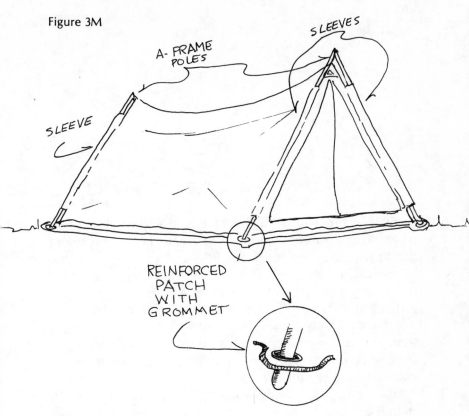

weight or stability (see Figure 3L). It is used by several manufacturers.

Instead of having the fabric nailed or tied to the frame, the legs of the modern A-frame pass through sleeves or pockets sewn into the tent fabric, and the bottom of the legs go through a grommet (or loop) set in a reinforced patch on the tent floor (see Figure 3M). This makes the structure self-supporting, although most campers guy it for even greater stability, and since the legs can't be forced into the earth too far, the tent doesn't have to be constantly adjusted.

Many backpacking tents and other small models employ an A-frame/I-pole combination with the A-frame in

Figure 3N

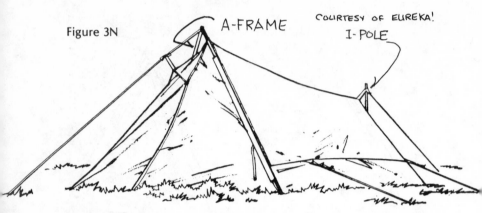

A-FRAME

COURTESY OF EUREKA!

I-POLE

front and the I-pole in the rear (see Figure 3N). Some firms produce models with an A-frame in the middle of the tent (either inside or outside) and use I-poles at the two ends (see Figure 3Q). When A-frames are used both front and rear (called a *double A-frame*), the tent is much more stable than the A-frame/I-pole combination.

Advantages: spills wind; self-standing; less expensive than the more exotic designs; easy to erect and disassemble, even at night; the triangular shape sheds water and even the wettest of snow.

Disadvantages: needs some guying, since it's not as stable as the models designed specifically for windy

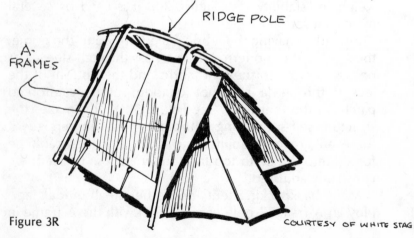

RIDGE POLE

A-FRAMES

Figure 3R

COURTESY OF WHITE STAG

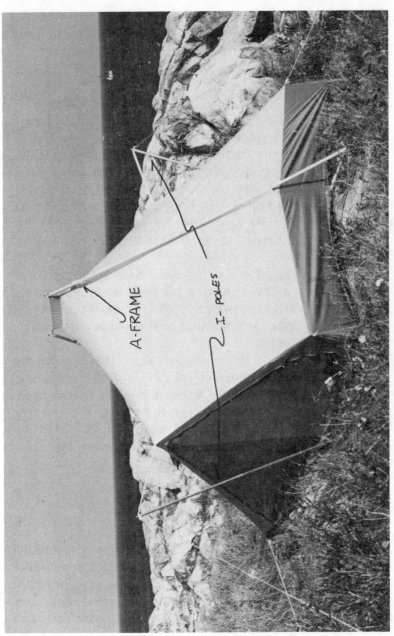

A-FRAME

I- POLES

Figure 3Q. Fortnight II. (Courtesy of Gerry)

conditions; more expensive than I-poles; the sleeves through which the legs pass can become frayed and tear; damaged poles are relatively costly to replace.

A-FRAME WITH RIDGE POLE

This is basically a double A-frame design that is connected along the ridge line by another pole (see Figure 3R). This allows the framework to be assembled quickly. The fabric is suspended from the ridge pole and legs of the two A-frames, making this design self-standing and able to be moved once it is set up without having to tear it apart. The rain fly can either be suspended from the ridge pole or draped over it and then guyed to the ground.

Advantages: portability and easy access through the entrance, since the structure seldom needs guy lines; sheds water and snow easily; roomy; self-supporting; comes with optional vestibules that provide extra space for extended tent stays.

Disadvantages: a little heavier than the regular A-frame; takes longer to erect and, like the dome frames, which are detailed a little later, it does have one odd disadvantage: if someone isn't in the tent or if it isn't full of equipment, a sudden gust of wind can lift the structure, causing it to "tumbleweed," or roll away, from the campsite. This is rare, but it can happen.

DRAW TITE

This unique design is also known as the straight bar exoskeleton or the Blanchard. It was invented some 20 years ago by Robert Blanchard, who holds seventeen different patents on its design, and is a favorite with expeditions and mountain climbers (see Figure 3S).

The Draw Tite features a frame that is under constant

Figure 3S

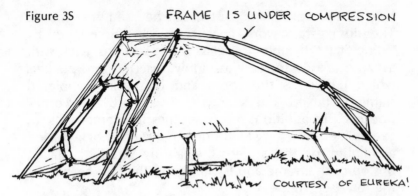

FRAME IS UNDER COMPRESSION

COURTESY OF EUREKA!

tension, thereby keeping the fabric tight and virtually wrinkle free. As explained by Eureka! Tent, Inc., the manufacturer of the Draw Tite: "The framework, of lifetime corrosion-proof, tempered aluminum alloy, is set up under compression and keeps the sewn-in floor spread flat and the tent drawn tight at all times, even when held off the ground. Elastic shock cords and simple brass hooks hold the tent snugly to the frame, absorbing wind shock and rain shrinkage. They reduce strain on the fabric to a minimum." The rain fly fits over the frame and is guyed to the ground.

Advantages: extremely wind stable; self-supporting, requiring no stakes or guys, so structure can be erected on any terrain from sand and mushy tropical floors to rocky, frozen ground and even ice; portable; good ventilation even in miserable weather; roomy; comes in sizes from one-man to six-man models.

Disadvantages: heavier than regular wind tents; frame segments are not shock corded, which can cause a problem in erection; takes a relatively longer time to assemble and take down; needs guying in extreme winds.

DOME

This tent is aerodynamically designed for winter and

heavy wind conditions, but it can be used year-round. The dome framework consists of, as pointed out by White Stag, "flexible fiberglass frame sections, each with a flared ferrule for easy assembly, inserted into pockets which criss-cross the room and then the completed frame is flexed and simply inserted into the corner pockets." In addition, most rain flies for dome tents, as exemplified by White Stag's design, have a shock-corded aluminum rod that stiffens the rain fly. The fly is placed on top of the frame and is clipped to the four corners of the tent (see Figure 3T).

Naturally, this type of framework screams for exotic designs, and there's no more creative a designer than Bill Moss of Camden, Maine. As pointed out by L. L. Bean, Inc., a long-time supplier of camping equipment, Moss's idea is that "primary emphasis on the functions that tents should perform—comfort, safety and economy of weight—plus a careful use of color results in tents that are also unusually attractive in appearance.

"Moss tent designs are based on tension. Construction is exceptionally strong and flexible, characterized by shapes that 'spill' wind and minimize drag. Flexible, lightweight fiberglass wands are used to impart a high degree of tensile strength to the walls, keeping them taut and eliminating flap. They are virtually unbreakable and completely self-supporting. The fabric itself thus becomes a structural part of the tent. No guylines are necessary. Erection is fast and uncomplicated. All of these dome or tunnel tents can be erected by one person in less than six minutes."

A good illustration of an exotic yet completely functional tent design is Moss's Sundance (see Figure 3U). There will be more about his ideas in later chapters.

Advantages: as much as fifty percent more room inside with the same floor area as I-poles or A-frames; self-standing and usually requiring no guying or staking;

Figure 3T

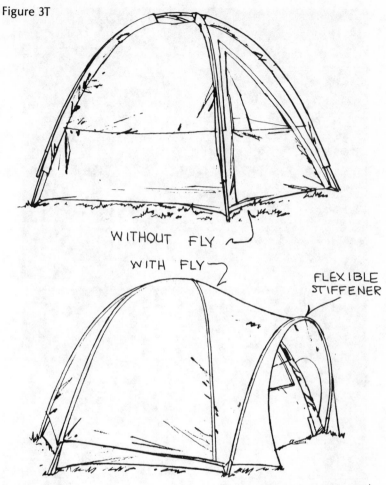

WITHOUT FLY

WITH FLY

FLEXIBLE STIFFENER

COURTESY OF EUREKA!

extremely stable in even heavy winds and because of their shapes, domes can experience the same type of "lift" that allows airplanes to fly, and, therefore, the tent wall is sucked out even more, providing more interior room.

Disadvantages: often heavier than other models; relatively difficult to erect; though the problem has been

Figure 3U. (Courtesy of Bill Moss)

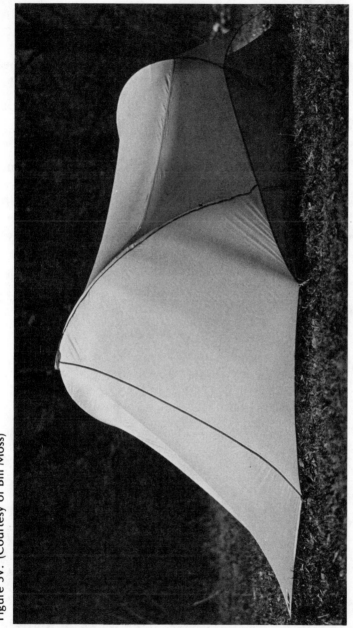

Figure 3V. (Courtesy of Bill Moss)

lessened by the use of fiberglass in place of aluminum, there is a tremendous amount of strain on the various joints, causing wear (This can be solved, however, by extra care in maintenance.); pole may flex in very heavy winds (but guying can eliminate this); ventilation often a problem.

TUNNEL

Basically the same frame design as the dome, the major difference between the two is that the tunnel frame allows a longer and roomier model with optional vestibules for even more space (see Figure 3V). The entrances offer one disadvantage in that the camper has to crawl in and out in the smaller models, but otherwise the tunnel and dome have the same advantages and disadvantages.

Figure 3X

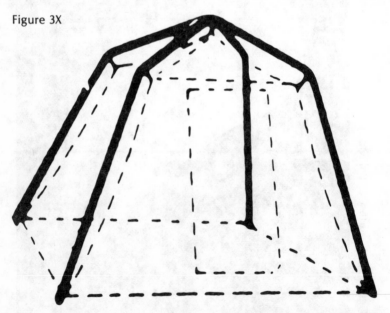

COURTESY OF WHITE STAG

Figure 3W

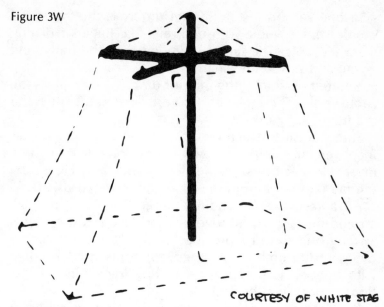

COURTESY OF WHITE STAG

UMBRELLA

This type of frame has an inside version (see Figure 3W) and an outside one (see Figure 3X). It is also called a skyhook frame. It is the larger version of a family shelter. It provides plenty of interior space and is self-standing.

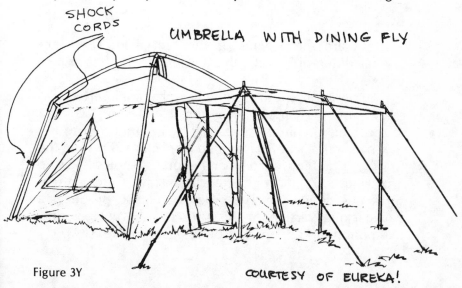

SHOCK CORDS

UMBRELLA WITH DINING FLY

Figure 3Y

COURTESY OF EUREKA!

In one version of the interior umbrella the frame is a simple center I-pole with umbrella-like fingers radiating from the pole. The fabric is draped over the frame and guyed and staked into shape.

Another version of the interior umbrella eliminates the center pole. The entire structure is locked together and the fabric hung over it.

Some outside structures have a center pole, but this inconvenient feature is slowly being done away with by most manufacturers. In the design shown in Figure 3Y, the fabric is hung from the frame by shock cords.

Advantages: self-standing; large enough to walk around in; easy to put together and take apart; optional awning offers even more space; rain fly is easily slipped over outside umbrella and guyed; some models come with a net enclosure, providing a bug-free dining/relaxation area outside the tent itself.

Disadvantages: large and heavy, often weighing up to sixty or seventy pounds; needs some guying in harsh weather; the outside awning or net enclosure can be a disaster in a sudden storm, blowing down and ripping part of the main tent if it isn't taken down immediately.

BOX

These are the biggies of the tent field. There are several different designs that fall under the heading of box tents.

One design has four flexible bars that are inserted into sleeves on the canopy (see step 1 of Figure 3Z). Two telescoping aluminum poles are then connected to the flex-bars in the sleeves (see step 2 of Figure 3Z). Then the tent is pushed into shape by telescoping I-poles (see step 3 of Figure 3Z) and tightened by staking. Coleman features this design in large models and also produces a backpacking version of this structure that weighs only five pounds and sleeps two people (see Figure 3AA).

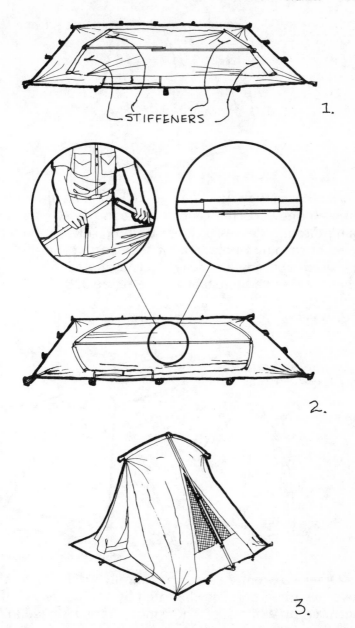

STIFFENERS

1.

2.

3.

Figure 3Z   COURTESY OF COLEMAN

Figure 3AA

COURTESY OF COLEMAN

Another design has a ridge bar running through a built-in sleeve on the canopy and two "eave" bars running through sleeves or loops on both sides of the tent. The telescoping I-poles are then connected and the tent is raised, one side at a time. A center pole is added and the tent is staked for stability (see Figure 3BB).

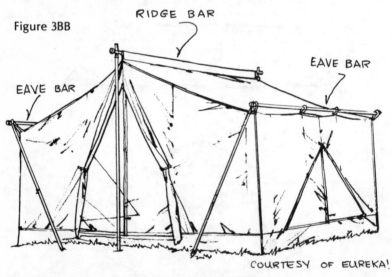

Figure 3BB

RIDGE BAR

EAVE BAR

EAVE BAR

COURTESY OF EUREKA!

Other box tent designs are basically variations on the two described here, as illustrated by two models manufactured by White Stag. These models are designed to be connected to a van/camper or pickup truck by magnetic tape and are self-standing. But the frame differs only in that the two upright I-poles criss-cross each other on the side (see Figure 3CC). Otherwise, the frame is much the same as the first one described.

Advantages: surprisingly stable, even in high winds; large and roomy enough to accommodate several campers comfortably, even if confined to the tent for a long period because of inclement weather; self-supporting, although guying and staking is recommended for safety's sake.

Disadvantages: so big and bulky that they have to be hauled to a campsite by vehicle; relatively difficult to put up and take down; tend to be the most expensive of the various frameworks.

VAN TENT

COURTESY OF WHITE STAG

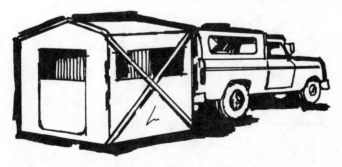

VAN/PICKUP TENT

Figure 3CC

INFLATABLE SUPPORTS

Inflatable supports are small, plastic tubes that are blown up by mouth and used to support a one-man emergency tent (see Figure 3DD).

Advantages: cheap; small enough to carry in a large jacket pocket; easy to put up.

Disadvantages: no stability; easily punctured, rendering the entire framework useless.

These are the basic support systems on the market at the time of this writing. By the time you read this, however, there may be another dozen or so completely different frameworks in camping supply stores. With the innovations in materials exploding on the scene, it's logical to assume that the more inventive tentmakers will soon be marketing totally different frames to hang the fabric on, as radically different as the dome structure differs from a buffalo hide draped over a couple of convenient tree branches.

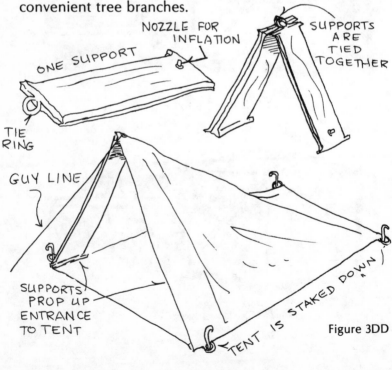

NOZZLE FOR INFLATION

ONE SUPPORT

SUPPORTS ARE TIED TOGETHER

TIE RING

GUY LINE

SUPPORTS PROP UP ENTRANCE TO TENT

TENT IS STAKED DOWN

Figure 3DD

4

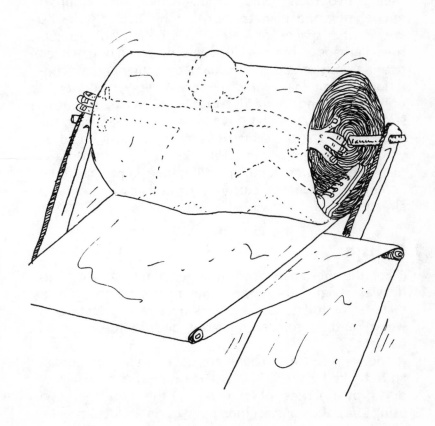

Fabrics

Like frames, tent fabrics have also evolved, with new materials revolutionizing the industry. They have progressed from animal skins to felt to rough, open weaves of hemp, cotton, and wool to close weaves like canvas and synthetic fibers. But it wasn't until after the Second World War, when nylon was introduced on the consumer market, that a real revolution took place. The new lightweight fabrics meant that tents large enough to shelter two men could be made light and compact enough for one man to carry on his back. And who knows what space-age material will be available tomorrow? Perhaps a thin but durable synthetic fabric, woven over a transistorized, self-contained heating and air conditioning unit that works off solar energy, and weighs only five pounds for a five-man tent? Maybe a tent so strong and so large that it could shield six men from the elements and at the same time so small that it could be carried in a shirt pocket? Who knows? But until these plausible fantasies shimmer to life, the camper has to be content with what is currently on the market, choices already plentiful and varied.

COTTON

For the most part used in larger tents because of its heavier weight, the cotton that makes up the canvas material of today's tents comes in a wide selection of weaves and weights, each geared toward a specific need or want.

The weaves, for instance, are duck, drill, twill, poplin, and combed cotton. The last is simply a material produced by a process of combing out the short fibers and using only the long strands for weaving. Duck, poplin, and combed cotton have a simple but the most effective weave pattern, with one strand of material going over one strand and then under the next (see Figure 4B). Twill

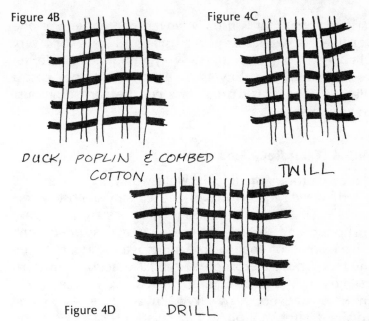

Figure 4B

Figure 4C

DUCK, POPLIN & COMBED COTTON

TWILL

Figure 4D   DRILL

is woven with a strand going over one and under three (see Figure 4C); drill has a strand over one and under two (see Figure 4D). The last two weave patterns are used with the coarser and cheaper cottons and aren't as durable as duck, poplin, or combed cotton because they're more susceptible to abrasion and the snags that trigger rips.

Other terms for tent canvas are Army duck, double-filled, single-filled, and pima cotton. The first three refer to material made of two strands of cotton twisted around each other and then usually woven one over and one under. The finished product is durable but also a little heavier and more expensive. Pima cotton (also known as Egyptian cotton) is a term you'll hear often and refers to a special cotton composed of long, silky, but extremely strong fibers. This material originated in Central and South America and is now presently grown in Egypt and the southwestern part of the United States. It is a high-grade material that is usually woven one under and one over like duck. Because it is so silky, it allows for a very tight weave, a factor that is very important but often overlooked by tent buyers. A tight weave does cost a little more, but it is worth it if a cotton-fabric tent satisfies your needs.

Natural Water Repellency (NWR)

Since cotton is organic, it swells when it gets wet, and in a tight weave, this makes the canopy almost waterproof. It will still leak, especially in a constant and heavy downpour, and should you brush up against the canopy or have some of your gear leaning against a part of the tent that isn't waterproofed, capillary action can start, getting you and your equipment soaking wet. Still, cotton enjoys a very high NWR. In addition, as will be explained later, cotton can be waterproofed or made even more repellent by chemicals. Unfortunately, this has its drawbacks, because it cuts down on the breathability factor.

Breathability Factor (BF)

The most difficult to judge and frequently ignored feature of a tent is whether it breathes or not. If you had to worry only about the outside moisture, a big Baggie would suffice. But the body also gives off moisture in the form of water vapor. The average man respires about a pint of this vapor during the night and up to a quart and a half if confined to a tent all day and all night. Multiply that by the number of people occupying the tent, and you can see why the inside atmosphere can become as muggy, wet, and uncomfortable as the outside. Another factor is that the temperature inside a tent is higher than outside, even on hot summer nights, and this means the water vapor will condense on the inside of the cooler canopy and saturate you and your equipment if it isn't allowed to escape. The reason for this is that the lighter-than-air water vapor is composed of molecules that are much smaller than raindrops, which rise to the canopy. If the fabric is *not* waterproofed, the vapor can pass through even the tightest of weaves and either dissipate in the open air or condense on the underside of the rain fly, where it drips harmlessly to the ground (see Figure 4E).

Figure 4E

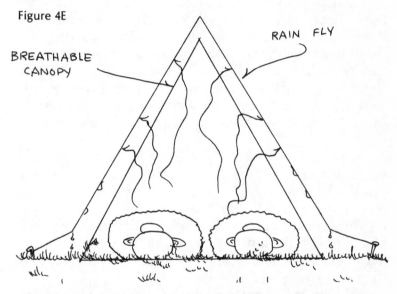

Cotton rates high on breathability. Even fabrics that are composed of eighty percent cotton and twenty percent synthetics like Dacron or rayon have high breathability factors.

Advantages: high NWR and BF; wide selection of weights and weaves, satisfying more varied needs and cost factors; costs less than synthetics.

Disadvantages: heavier than synthetics when dry and many times heavier when wet; can rot and mildew; can't be packed wet, so must be thoroughly dry before a camper can move on; tears more easily than the better grades of nylon; requires much more maintenance and care than synthetics. Best considered for the larger tents that will be hauled in by vehicle.

SYNTHETICS

Plastic

In addition to tarp tents (see Chapter 8, *Other Shelters*), there are several kinds of tents made entirely of waterproof plastic. Chapter 6, *Tent Classifications,* will detail these tents more fully, but as should be clear from the preceding discussion of breathability, these tents shouldn't be considered as anything but emergency shelters. In a plastic tent, you can become as wet and miserable as if you didn't have anything to ward off the raindrops.

Discomfort isn't the only reason for not recommending plastic tents except as emergency protections—they can also be dangerous. For example, the ends of the tube tent (see Chapter 6, *Tent Classifications*) can be sealed off with tape, and this is much like putting your head in a plastic food storage bag and taping it tightly around your neck. Campers have actually suffocated in plastic tents that were completely sealed.

Advantages: cheap; easy to pack; some models are disposable; lightweight; high NWR.

Disadvantages: virtually no BF at all; the fabric can rip and be damaged more easily than cotton or woven synthetics; uncomfortable for any length of time; can be dangerous and even lethal.

Nylon

Nylon is the most popular material for backpacking-size tents because it is strong, durable, lightweight, less apt to rip and tear than cotton, and requires less maintenance. Nylon can be woven into extremely tight weaves, and this helps add to its natural water repellency without sacrificing breathability. It is also highly resistant to wind.

Nylon comes in two different weaves: taffeta and ripstop. Taffeta is woven like duck, a strand over one strand and then under another, and is very smooth and highly resistant to abrasions. Ripstop is also woven like duck, but every one-eighth to one-quarter of an inch (depending on the manufacturer) a special thread of two fibers twisted around each other is woven into the pattern along both the weft and warp of the material (see Figure 4F). Ripstop does just what its name implies; stop rips from becoming too large. However, it is more prone to abrasive damage than the smoother taffeta. Ripstop is

Figure 4F

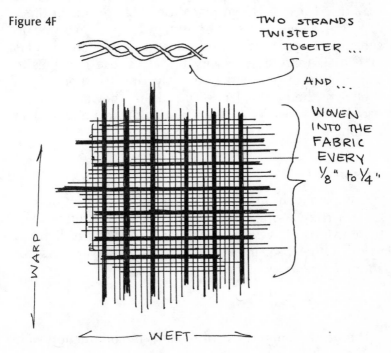

TWO STRANDS TWISTED TOGETER ...

AND ...

WOVEN INTO THE FABRIC EVERY $\frac{1}{8}$" to $\frac{1}{4}$"

WARP

WEFT

lighter but slightly more expensive. Still, it's worth the cost if you're going camping in an area where the terrain can cause damage to a tent or if you're a dedicated klutz. You can easily spot the difference between the two weaves, because taffeta is silky smooth with an unbroken texture while ripstop has a rectangular pattern in its weave because of the extra fibers.

Advantages: BF as good as cotton, but a little lower in NWR because it doesn't swell when wet; cheaper and easier to care for than cotton; lightweight; highly resistant to rot and mildew, so it can be packed wet if the need arises; doesn't tear as easily as cotton; resistant to abrasions.

Disadvantages: resists water-repellent chemicals one might desire to add to increase the tent's NWR; stretches more than cotton, so usually a nylon tent has to be readjusted if it is left in place for long periods of time; can lose some of its fiber strength when exposed to sunlight over a long period (most experienced campers take their tents down every few days, use rain flies to shield them during part of the day, or pitch them in the shelter of the natural landscape so that they aren't bombarded by sun rays all the time); more expensive than cotton.

Dacron

Comparatively new on the market, some tentmakers use Dacron canopies because this material stretches less than nylon. However, it isn't as strong or as light as nylon. Otherwise, its advantages and disadvantages are about the same as nylon.

Rayon

The most common use of rayon is in conjunction with

nylon. The two materials are laminated to form the newest tent material, Gore-Tex.

Gore-Tex

As explained by White Stag, Gore-Tex is a "product of W. L. Gore & Associates, Inc., and is a new microporous membrane which has such tiny pores (9 billion per square inch) that liquid water cannot pass through it. However, water vapor can pass through it, so evaporated perspiration can escape from the inside of the tent instead of collecting in the form of condensation."

Since Gore-Tex is only .001 inch thick, it has to be placed between layers of other materials to form the tent canopy. In most designs, ripstop and rayon or nexus polyester form the outer and inner layers, with the membrane in the middle.

Advantages: high in NWR and BF; virtually waterproof, so a rain fly isn't needed.

Disadvantages: adds a little more weight to a tent, but this is offset by not having to carry a rain fly; in cold weather, vapor tends to frost on the fabric a little more than on a cotton frost liner or nylon canopy; expensive, costing almost twice as much as nylon.

Figure 4G

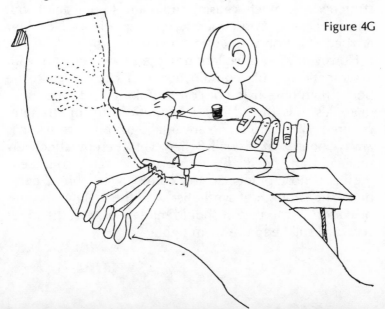

THREADS, SEAMS, AND STITCHING

You could have the best frame made out of the finest material available; you could obtain the newest fabric and have it all put together by a genius; and yet if the seams, the threads, and the method of stitching were inferior, the tent would be a bummer. Fortunately for the serious tent buyer, the quality tentmakers take great pains with these aspects. Just make sure you take advantage of their creative thinking and examine the seams, threads, and stitching before you slip the credit card on the counter or sign the check.

Threads

A general rule that used to apply here is that synthetic threads should be used with synthetic fabrics and cotton should be used with cotton. The reason for this was that an owner of a nylon tent sewn with cotton thread sometimes found the seams tearing because the thread had rotted, something he didn't have to worry about with a synthetic fabric. Polyester with cotton threads didn't work either, because the polyester has a tendency to stretch and, with the heavier cotton, it occasionally stretched too much, causing the seams to pull apart and, naturally, leak. Consequently, the experienced camper advised, "Cotton with cotton; nylon with nylon."

However, innovative tent designers have come up with a synthetic/organic combination that has proved to be better than pure nylon or cotton. Taking the best of both materials, this thread has a synthetic core for strength, covered with cotton. The advantage of cotton thread has always been that it swells when wet, thereby effectively sealing off the needle holes, preventing water from seeping in at those points. So this cotton/nylon thread combines the strength of synthetics with the sealing qualities of cotton. If the cotton should mildew or tear, the inner core will still keep the seam tight.

Seams

In a quality tent, the seams are as scientifically designed as the rest of the shelter. Similarly, in an inferior model they're as ignored as the rest of the tent.

One seam to avoid is the plain seam (see Figure 4H), where the two pieces of material are merely joined by

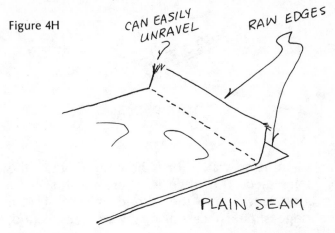

Figure 4H

CAN EASILY UNRAVEL

RAW EDGES

PLAIN SEAM

thread and the edges left exposed. This seam has a tendency to unravel and tear because the raw edges aren't folded over as they are in the flat, or lap fell, seam (see Figure 4I). This seam is popular because the edges

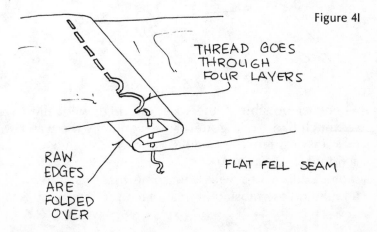

Figure 4I

THREAD GOES THROUGH FOUR LAYERS

RAW EDGES ARE FOLDED OVER

FLAT FELL SEAM

are folded in and around each other, allowing the thread to go through all four layers and effectively cutting down on fabric deterioration at the seams. Another quality seam is the bound seam (see Figure 4J), in which another piece of material is sewn to the fabric, covering the raw edges.

Figure 4J

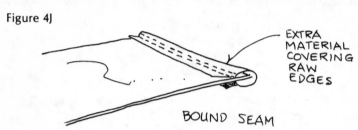

EXTRA
MATERIAL
COVERING
RAW
EDGES

BOUND SEAM

Another feature to watch for is how the ridge of the material is handled. The ridge is where the first layer overlaps and is sewn onto the second one (see Figure 4K). The ridge should face down, toward the ground,

Figure 4K

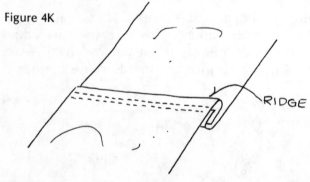

RIDGE

whenever possible. Otherwise, the ridge will collect water much like a rain gutter, and eventually the water will leak in through the needle holes or through capillary action (see Figure 4L).

Since all seams will leak somewhat because of the needle holes, most tent manufacturers supply a seam

Figure 4L

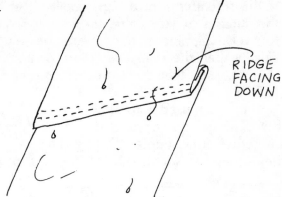

RIDGE
FACING
DOWN

sealant. This should be applied to both sides of the seam so that it can bond together through the holes. You should also carry an extra bottle of seam sealer with you to patch up small leaks that will occur from time to time. If desperate, you can always use melted wax from a candle to seal off the holes (see Figure 4M).

Figure 4M

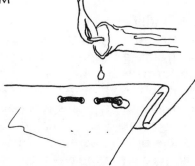

Stitching

In tents, more stitching isn't necessarily better. Many stitches mean a lot of needle holes, and this means the material is weakened at that point, much like perforated paper. Therefore, the best method of determining whether the stitching is of a high quality is to actually count the stitches. If you come up with an average of seven to twelve per inch, chances are the tent is of a better quality than the ones with more or less stitches. Remember, though, that this is just one method, and it isn't infallible.

The kind of stitch used is also important. It should be a lock stitch, where the upper and lower threads are locked together at close intervals (see Figure 4N). This is the strongest type of stitch, and it is less apt to unravel.

Figure 4N   LOCK STITCH

Also check the beginning and end of each seam to make sure the stitching is back-stitched or bar tacked, as this eliminates the possibility of the thread unraveling.

REINFORCEMENTS

Since a tent fabric is under constant tension, it should be reinforced at various stress points. Otherwise, the fabric may break away from a guy line or stake, usually in a miserably cold rainstorm with biting winds, possibly ruining your equipment and the tent itself.

Figure 4P illustrates where the typical stress points of a tent are located, although these areas may differ from model to model. To be on the safe side, make sure there's extra material (usually a heavy-duty nylon or nylon webbing, like that used in backpacks) sewn onto the tent fabric wherever there's a point of tension, like where shock cords are attached (see Figure 4Q), where guy lines are connected (see Figure 4R), or where stake loops and grommets are attached (see Figure 4S). Additional points to check are the peaks, corners, and pole

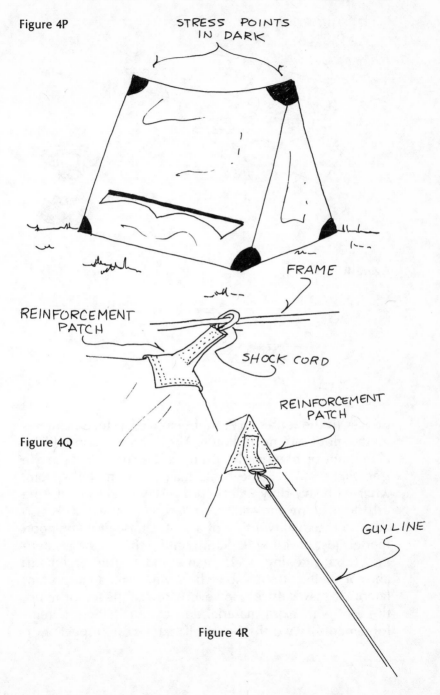

Figure 4P

STRESS POINTS
IN DARK

FRAME

REINFORCEMENT
PATCH

SHOCK CORD

Figure 4Q

REINFORCEMENT
PATCH

GUY LINE

Figure 4R

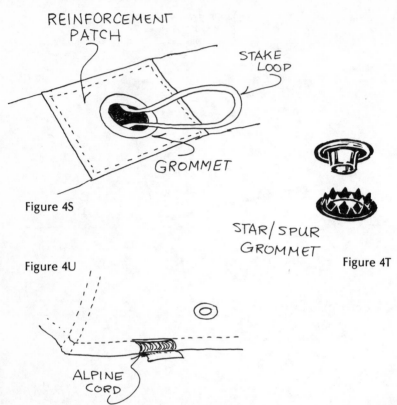

REINFORCEMENT PATCH

STAKE LOOP

GROMMET

Figure 4S

STAR/SPUR GROMMET

Figure 4U

Figure 4T

ALPINE CORD

sleeves (if the tent is an A-frame) as well as the grommets in the floor or the pullouts. The grommets should be aluminum or brass and should be the spur, or star, style (see Figure 4T). Some tents have a perimeter system, where a heavy-duty nylon cord is sewn right in with the tub floor along the stake line (see Figure 4U). This also adds to the quality rating of a tent, although many good models just have regular cloth reinforcing or extra stitching. Extra stitching is also a method of reinforcement, and for shelters to be used where wind isn't an important factor, it may suffice. However, for real peace of mind, the tent with extra material *and* extra stitching as reinforcements is worth the small extra cost, especially if these features save you the trouble of having to chase your tent down a dark ravine some miserably rainy night.

ZIPPERS

In the early days of tenting, shelters were closed with ties, loops of canvas, or leather thongs. These fasteners kept out some of the rain, but they didn't keep out the wind or the cold or the bugs. Then steel zippers made their appearance, although most quality tents kept the ties as well. Why? Because the zippers only worked a fraction of the time. They jammed, they broke, they froze, and sometimes they just didn't work for reasons known only to the Great Zipper God. In addition, they rusted, and the corrosion often caused the tent fabric to rot. Consequently, the ties were backup systems that were used as often as the zippers.

Improvements continued, however, and brass zippers began replacing steel ones. Although not as temperamental as the steelies, brass zippers were finky at times, usually in an emergency situation when the camper wanted to get into or out of the tent in a hurry. In addition, the brass still corroded, triggering the chemical reaction that rotted the material. Then the nylon zipper was perfected, and campers all over the world sighed in relief.

There are two types of nylon zippers, the coil and the large-toothed zipper. The coil has no teeth and is used in places on the tent where the zipper has to go around curves or angles. Where the zipper runs in a straight line, the large-toothed type is normally installed. Whether your tent has coil or large-toothed zippers, be sure they are the double slider design, the kind that can be opened at either end. These help in controlling the ventilation, as you'll be able to open the bottom of the door or the top, depending on the weather, and they are easier to operate in even the darkest of nights.

VELCRO

Velcro is that familiar closing device, made up of tiny nylon hooks and loops, that has grown so popular in clothing. It was utilized on tents as a replacement for zippers for a while. But it proved to be too easy for the wind to blow open, and tent designers went back to zippers. However, some models feature both zippers and Velcro, with the hook-and-loop closure acting as a backup system.

RAIN FLY

As already mentioned, the ideal tent fabric is the kind that allows the moisture you give off to seep through the canopy at the same time it keeps the raindrops from coming in. Only one tent, the Gore-Tex, really fulfills both these requirements without a rain fly. However, it is not necessarily better to dispense with a rain fly, as the fly does more than just keep the rain off the canopy while allowing it to breathe.

For one thing, when the tent is properly rigged with six or seven inches of space between the fly and the canopy, an air pocket is created, forming the double-wall construction. Air is trapped in this space and, since still air is one of the best insulators known, the tent is cooler in summer and warmer in winter because of that layer of unmoving air caught between the two walls. The rain fly also serves as a shield from the harsh rays of the sun, which can weaken nylon over a long period of time. In addition, the fly can be carried by itself and used as an emergency shelter. A rain fly won't provide protection from bugs, but it will shed rain well enough to keep the camper fairly dry until a sudden shower is over (see Chapter 8, *Other Shelters,* for more information).

TUB, OR BOAT, FLOORS

This is a feature that just about every tent manufacturer puts in most models, and there's no reason why a camper should buy a shelter without a tub, or boat, floor.

Basically, this type of floor consists of heavy-duty nylon

Figure 4V

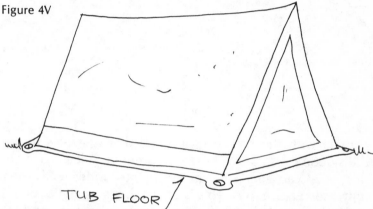

TUB FLOOR

or Army duck that has been completely waterproofed and sewn right to the rest of the tent fabric, often running six or seven inches up the sides (see Figure 4V). This sill helps keep the water that runs off the rain fly from leaking in as it splatters on the ground. On some tents designed for rainy or tropical climates the sill runs even higher up the side of the tent.

The tub floor not only keeps water out but it also provides an excellent barrier to crawling vermin and other creepies that can and do get inside tents without tub floors. It takes only one experience of sliding into a sleeping bag only to discover it's full of five-inch slimy slugs to be convinced of the value of a tub-floor. Why suffer through such an unnerving experience when you don't have to?

However, there can be a condensation problem with tub floors. This is caused by the warmer floor making

Figure 4W

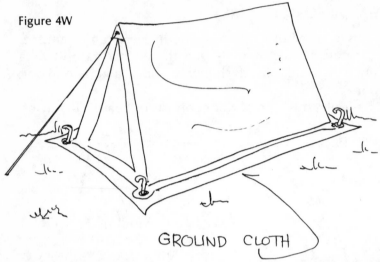

GROUND CLOTH

contact with the colder ground so that it sweats under the sleeping bag and where equipment is piled. To avoid this, most campers use a ground cloth underneath their tents (see Figure 4W). This is an inexpensive, lightweight, and waterproof sheet of plastic. It retards floor condensation, because all the moisture that forms will be on the underside of the ground cloth, which is next to the earth. Since it's waterproof, the ground cloth prevents the condensation from seeping through to the floor of the tent.

Sometimes there is also a problem with some leaching of ground-level moisture through the heavy threads used in sewing the tub floor. A good plastic sealant will solve most of this. Make sure that the tub floor in the tent you buy is a wraparound, with few, if any, seams near the ground. Finally, always check out a tub floor to make sure most of its seams are higher up, thus avoiding these potential trouble spots.

Don't forget, though, that the floor is still made out of a fabric and it can rip, tear, and suffer from abrasions. So be careful in picking a spot to pitch the tent, avoiding areas where there are many sharp pebbles and stones, which can damage the floor. A ground cloth will help prevent this abrasive harm by acting as a shield for most of the forest debris. In addition to checking the site thoroughly, take off hiking boots *before* entering the tent whenever feasible or take them off as soon as possible upon entering. Some campers solve this problem by packing along a pair of "camp shoes," usually a pair of comfortable, broken-in sneakers. These not only cause less wear and tear on the floor but they also give your feet a rest from the heavy boots. Finally, if the floor doesn't have a cookhole, be careful if you cook inside the tent. The fuel from the stove—and from lanterns—can damage the floor. It can also puddle instead of being absorbed, since the floor is waterproof, and some of your equipment can soak up these material-rotting chemicals.

It should be obvious by now that a tub floor should be a feature of the tent you buy. The small savings realized by buying a tent without a tub floor just aren't worth giving up the convenience and safety the tub floor offers.

WATERPROOFING

No matter how tightly the material is woven, neither cotton nor nylon is completely waterproof. The fabric will leak eventually if a rain fly isn't employed or if the rain is a long, steady one. However, you can waterproof both types—if you really want to.

Some experienced campers swear by waterproofed tents and wouldn't use anything else. But when you compare the benefits of a waterproofed tent with the ones offered by the double-wall constructions, double wall far outstrip waterproofed tents.

For instance, one advantage of the waterproofed tent

is that the fabric is strengthened by the chemicals and therefore resists tearing better than untreated fabrics. But this benefit is offset by the fact that the waterproofed fabric doesn't breathe, and the comfort provided by a double-wall or Gore-Tex is far more important than the added strength of waterproofed material.

Pro-waterproofers argue that one doesn't have to mess around with rigging and carrying a rain fly if the fabric is treated, but these are weak points. First of all, it takes only a minute or two to set up most rain flies, so the time factor doesn't count for much. But more important is the fact that waterproofing a fabric adds weight, sometimes as much as doubling the tent weight. Therefore, it doesn't make much sense to treat a tent so that you don't have to carry a pound-and-a-quarter rain fly if you have to haul around a tent that originally weighed six pounds but now tips the scales at twelve pounds.

Still, if you want to purchase a waterproof tent, there are many different models available. If you want to treat the one you already own, most manufacturers sell the chemicals through catalogs or camping supply stores.

FIRE RETARDATION

This is a comparatively new idea in camping and, actually, it's really not warranted. Legislation forcing manufacturers to coat most tents with fire retardant chemicals (also known as FR agents) that comply with the standards set by the Canvas Products Association International (CPAI-84) came about mostly because of a few unfortunate incidents involving children playing in backyard tents. For the most part, the shelters involved in these incidents were made of wax-treated cotton (an old method of waterproofing cotton) and were meant for pitching in the backyard, for play only. Some children were burned when the tents caught fire, and this triggered an avalanche of scare stories on how millions of people were sleeping in tents that self-destructed if they were pitched within sixty feet of a spark. Politicians began running down the publicity trail, and laws were passed forcing tent manufacturers to treat fabrics with fire retardant chemicals.

Luckily, these chemicals don't harm the fabric or add any noticeable weight. But the treatment does add to the cost, sometimes as much as ten percent. It's just unfortunate that these laws came into effect, since campers sleeping in professionally designed and constructed trail tents are quite safe from fire if they exert reasonable caution.

A FEW MORE WAYS TO DETERMINE QUALITY IN FABRICS

Many hints on determining fabric quality have already been mentioned. Two more characteristics to consider are fabric weight and thread count. The former is usually expressed in ounces per square yard (for example, 1.9 oz./sq. yd.), and this figure is used in conjunction with the thread count to determine quality.

The thread count tells how many strands are in a square inch. It is usually expressed in the form 150 × 150, or just 150. This means that the fabric contains 150 threads per square inch running lengthwise and 150 widthwise. If the thread count is high and the weight per square yard low, the fabric is a tightly woven one that is stronger, more water repellent, and has a better breathability factor than one with a low thread count and high weight.

Usually weight will be included as part of the sales information, but the thread counts may not be listed. However, most manufacturers will supply thread counts upon request, and this information will help you make an intelligent decision.

Another way to weed out poorly made tents from good ones is to check the fabric to make sure there are no raw edges showing anywhere, at the seams or in other places. You should also check nylon tents to make sure the fabric has been hot cut. This is a method, used by most manufacturers, that seals the edges of the nylon with heat and can be recognized by the strands' bulby ends.

Naturally, one way to determine quality if you're in doubt is to ask experienced campers, people whose opinions you trust, what they prefer in fabrics and why. Then sort through those word-of-mouth recommendations and select the ones that fit into the framework of your needs. As already mentioned, there's such a variety of tents on the market that just about any desire can be satisfied.

5

Accessories

When people spend time in any kind of dwelling, permanent or temporary, they tend to improve or personalize the shelter. Tents are no different. Some of the tent accessories available make the soft home more comfortable, some make it more functional and versatile, and some are just luxurious items that aren't really necessary. This last category will be ignored here, since the plush extras are usually a matter of individual taste and listing them all would be a waste of time for most people. Instead, this section will concentrate on the supplementary gear that makes the tent more livable or functional.

For instance, although the vast majority of quality tent manufacturers include stakes along with their tents, a few models may not be so equipped, and you will have to purchase these small but necessary items separately. Or you may not like the type of stake furnished with the tent and prefer different designs. Or a stake may split or get lost and you will have to replace it. Or you just may want to expand your outdoor activities and therefore have to purchase the type of stakes created for a terrain different from the one you had been camping in. Whatever the reason, you may want to purchase a set of stakes at some time, and, since there are so many dissimilar styles, you should first determine your needs and desires, even for such a small item.

The following is a list of accessories that fulfill the more important functions of a tent—the items that make the shelter more comfortable or help adapt it for different environments and terrains or that are replacements for damaged or lost gear. If you want to indulge in the nonessential paraphernalia, it's up to you to ferret out what you want.

STAKES

Also known as pegs, tent stakes used to be made of

wood. Wood stakes often split, became soft and mushy from being in the damp ground, and had all the other disadvantages of wood. Now stakes are made of high-impact, nearly unbreakable plastic; rust-resistant steel; or tempered, corrosion-proof aluminum. The designs have evolved from the primitive tapered wooden peg with a notched top for the guy line to U-shaped stakes; tempered, lightweight models with hooks for guy lines; T-shapes; and skewers that are literally screwed into the ground. There are even stakes specially designed for use in snow and sand and stakes for rocky terrains or ice, where pounding something into the earth is almost impossible.

One word, though, about staking: many campers feel that since they have self-supporting tents, they don't have to bother with stakes. Often they don't. But to be on the safe side, to really enjoy the security of a stable tent, stakes should be used even with the self-standing designs, especially if there's a threat of stormy weather.

Plastic

There are two basic types of plastic tent stake: the hook, or T-stake, and the nail (see Figure 5B). Made of molded, high-impact plastic, these pegs are almost indestructible, are easy to pound into even semifrozen earth, and are fairly easy to extract. They usually come in bright yellow or international orange so that they can be quickly spotted in the forest duff. Plastic stakes are so lightweight (one-quarter to one ounce each) that the backpacker doesn't have to sacrifice any other gear to carry them. They are also inexpensive (approximately twenty cents for a nine-inch stake to thirty-five cents for a foot-long model). The plastic pegs are also weather resistant, impervious to rust or corrosion, can take heavy pounding, and are easily replaced, as just about every camping store carries them.

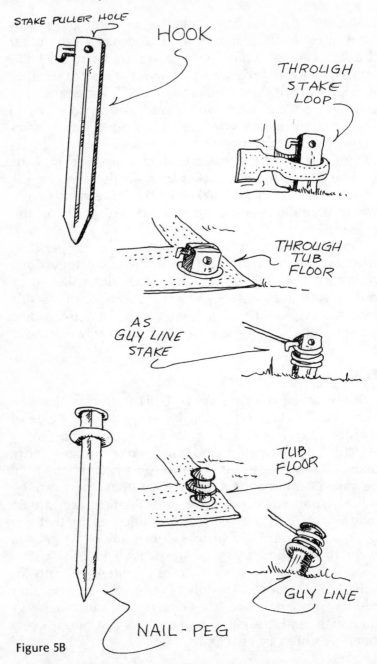

STAKE PULLER HOLE

HOOK

THROUGH STAKE LOOP

THROUGH TUB FLOOR

AS GUY LINE STAKE

TUB FLOOR

GUY LINE

NAIL-PEG

Figure 5B

Aluminum

The aluminum stake also offers a variety of designs. They come in the half-tube style (see Figure 5C), with a rolled top to hold the cord; in the "U," or staple, shape

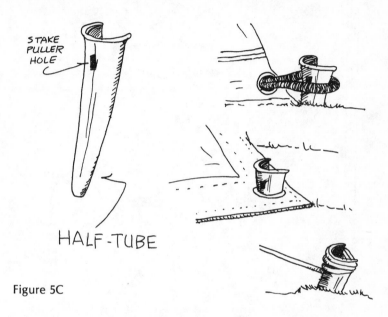

HALF-TUBE

Figure 5C

(see Figure 5D); and in the snow stake style, which is really a half-tube with four to six holes in it. The holes serve the following purpose: After the stake is pushed into the snow, its residual heat will melt the snow around it, with the water seeping into the holes; and once the stake cools off, the water will freeze around the peg and through the holes, anchoring it more securely (see Figure 5E). Also made out of aluminum are the skewer, anchor, and spike designs. These will be covered more fully in the following section on steel stakes.

The advantages of the half-tube and the snow stake are, naturally, their light weight (three-quarters of an ounce to one ounce each) and the ease with which they

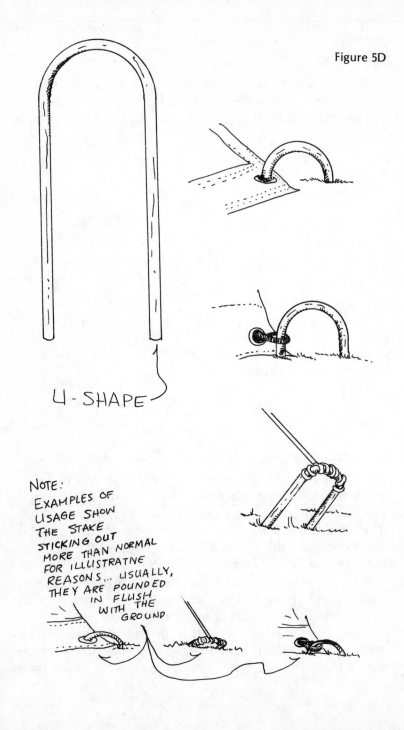

Figure 5D

U-SHAPE

NOTE:
EXAMPLES OF
USAGE SHOW
THE STAKE
STICKING OUT
MORE THAN NORMAL
FOR ILLUSTRATIVE
REASONS... USUALLY,
THEY ARE POUNDED
IN FLUSH
WITH THE
GROUND

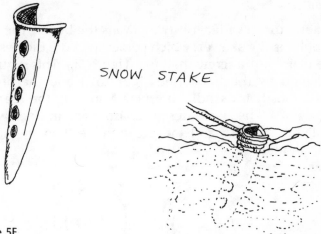

SNOW STAKE

Figure 5E

are packed, since they fit together in a compact bundle that can be tied with string or held together by a rubber band. They are also easily obtainable, rustproof, and corrosion resistant. The big disadvantage to half-tube or snow stakes is that they are best for soft, mushy ground or snow, tending to buckle or bend if pounded into hard terrain.

The U-shape, also known as the U-skewer, has the same advantages of the half-tube but not the big disadvantage, as it is able to take a heavier pounding into hard soil. Its shape also prevents the guy line from slipping off, no matter how harsh the weather. Aluminum U-shapes (there's also a steel staple) weigh about half an ounce each and cost about twenty-five cents each.

The aluminum skewer weighs one-third of an ounce and costs about fifteen cents each. The anchor is heavier, at an ounce per stake, and costs sixty cents per peg. The spike is the heaviest, weighing usually an ounce and a half to two ounces and will run from sixty-five cents for the nine-inch stake to ninety-five cents for the twelve-inch size.

Steel

Here, too, a variety of styles awaits the buyer. The most popular is the skewer, which comes in two basic designs: the plain or the screw shaft (see Figure 5F). This is quickly followed by the hook (see Figure 5G) and the anchor (used mostly for sand). The spike brings up the rear (see Figure 5H). The staple design is also made in steel, which is slightly heavier and more expensive than the aluminum model.

Figure 5F

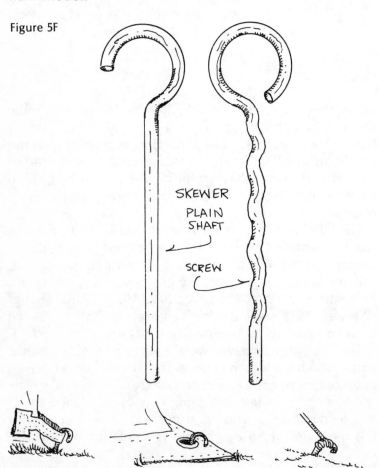

SKEWER
PLAIN
SHAFT

SCREW

HOOK

ANCHOR

SPIKE

Figure 5H

The advantages of the skewer, both aluminum and steel, is that it can be used in mushy soil, snow, and hard and rocky terrains. It can even be used on ice or on ground too hard to pound into by tying the guy line to the stake and piling rocks or ice on top of it to hold it fast (see Figure 51). Because of the eyehole or hook top, there's less chance of a guy line slipping off even in the

Figure 51

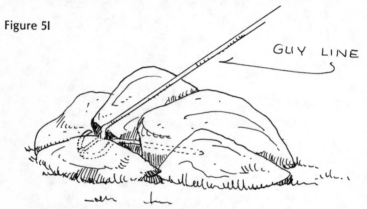

GUY LINE

heaviest of winds. Even the steel is lightweight, a package of eight stakes weighing five ounces. The eight-stake package costs about $1.50.

The disadvantage of the steel skewers, as compared to the aluminum one, is that the steel will begin to rust after a season and has to be kept constantly oiled to prevent pitting and weakening. Its big advantage is that it can take more abuse than the aluminum model.

The hook (weighing one quarter of a pound for a twelve-inch stake and costing about fifty cents each), the anchor (weight: one-quarter of a pound for an eight-inch stake; cost: eighty-five cents each), and the spike (weight: up to six ounces for the twelve-incher; cost: about sixty cents each) also have steel's disadvantage of being susceptible to rust, but all are good for rocky terrains where staking is difficult.

***

No matter what material the stake is made of, there are a few ways to check the quality of the design and manufacture. With metal stakes, be sure they are free from burrs or sharp edges that can cut a tent fabric or wear away a guy line. With the aluminum designs, try bending them in your hands. If they buckle, don't buy them. Examine the plastic stake closely to be sure there are no manufacturing cracks or splinters at the ends.

***

Accessories that are related to this category include rubber or soft plastic stake hammers, some of which feature a stake puller on one end (see Figure 5J); a plain stake puller (see Figure 5K); and a stake bag in which all the pegs and the puller or hammer can be carried with less chance of losing them.

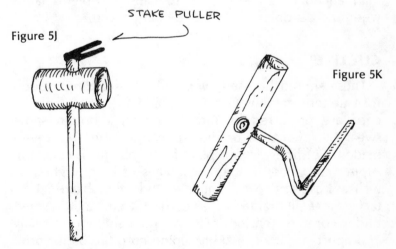

STAKE PULLER

Figure 5J

Figure 5K

Do-It-Yourself Stakes

In an emergency, you can always whittle your own stakes out of deadwood found on the forest floor (see Figure 5L). If snow conditions present a problem, you can use what are known as deadman anchors. These are

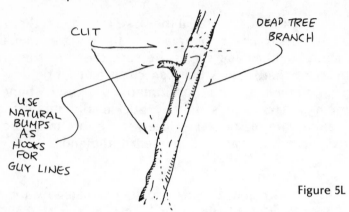

CUT

DEAD TREE BRANCH

USE NATURAL BUMPS AS HOOKS FOR GUY LINES

Figure 5L

plastic T-shaped pegs, tree branches, dead bushes, a stake bag, or anything else you can find to attach a guy line to and bury in the snow for security (see Figure 5M). Avoid using the objects as shown in Figure 5N for deadman anchors, as these tend to cause problems later, when the deadmen aren't needed any more.

GUY LINES

These should be used even with self-standing models for true tent stability. Guy lines made of $\frac{3}{16}''$ or $\frac{1}{8}''$ alpine cord are preferred by most campers, although some swear by $\frac{1}{4}''$ manila or sisal rope (the latter is a hemp made up of fibers from a plant found mainly in Yucatan). Alpine cord has the big advantages of being lighter as well as resistant to rot, mildew, and other bacterial attacks. Since all braided ropes tend to stretch a little when under constant tension, nylon's stretchability isn't really an important factor, making alpine cord the big favorite for guy lines. However, avoid using shock cord for the entire guy line, as this may prove to be too elastic. Quarter-inch rope guys as long as thirteen feet can be purchased for as little as eighty cents and for $1.25 you can get the same length complete with a plastic adjuster already attached and the loops wired together. Alpine

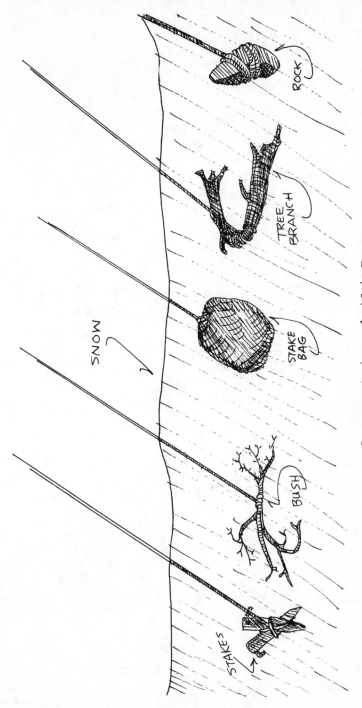

SNOW

ROCK

TREE BRANCH

STAKE BAG

BUSH

STAKES

DEAD MAN ANCHORS

Figure 5M

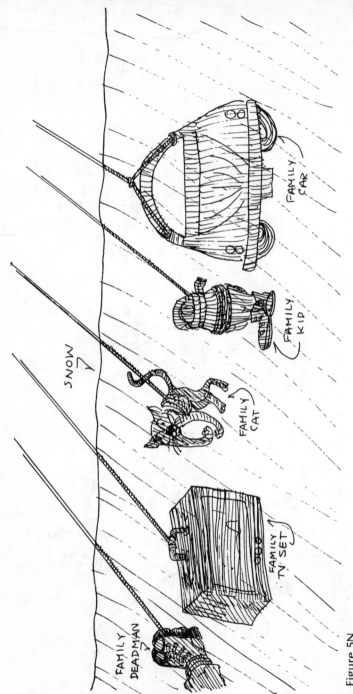

SNOW

FAMILY DEADMAN

FAMILY TV SET

FAMILY CAT

FAMILY KID

FAMILY CAR

Figure 5N

cord runs about three cents to five cents per running foot, depending on the size, but compared to the thirteen-foot guy line, alpine would cost only thirty-nine cents or so for the same length. Add that to alpine cord's lighter weight (a fifty-foot coil weighs only .16 lb.), and it becomes the better buy.

However, since all ropes tend to stretch, they have to be readjusted and tightened from time to time. This brings the next category to the surface . . .

CORD TIGHTENERS

These are small appliances for tightening guy lines. Figure 5P shows the various types available. However, it should be noted that there may be a brand-new concept on the market by the time you read this, since tents and their component parts are constantly evolving. Still, the cord tighteners currently on the market will be around for many years, and there are plenty of different models to choose from, each fitting some need and into some budget.

Cord tighteners are found in a wide range of prices: five cents for the rope slide; twenty-five cents for the tension, figure-eight tightener; thirty-five cents for a package of eight plastic slips; eight-five cents for a tension spring; ninety-five cents for a turnbuckle.

The style you prefer will depend on your experience. Most campers usually start off with the rope slide or slips and graduate to the more complicated tighteners as they become more adept at rigging a tent. Eventually, you'll start experimenting to find the type you like best.

FABRIC TIGHTENERS

The tent you pick may not have pullouts attached, but this doesn't mean you can't enjoy the tautness the pull-

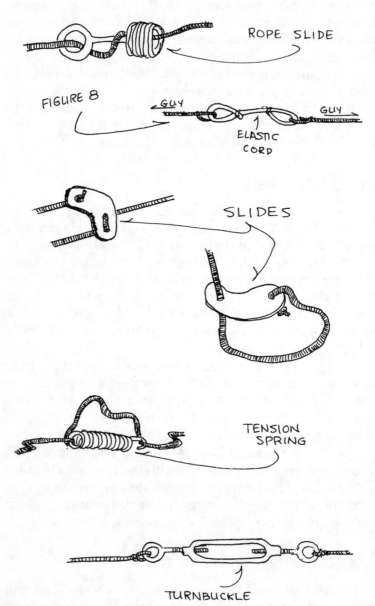

ROPE SLIDE

FIGURE 8

← GUY   GUY →

ELASTIC CORD

SLIDES

TENSION SPRING

TURNBUCKLE

Figure 5P

outs offer. By using either a vise clamp or a versa tie (see Figure 5Q) you can stretch out the sides of a loose tent, making them taut and keeping them from flapping in high winds. These fabric tighteners can also be used on emergency shelters or tarp tents for added protection and ease in rigging (see Figure 5R). These clamps have the added features of being lightweight and inexpensive, with a package of twenty selling for around $2.50.

Figure 5Q

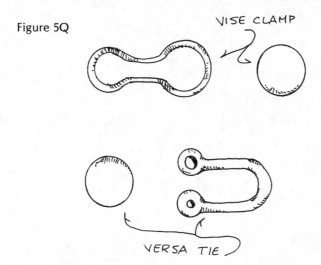

GROUND CLOTHS

These can be heavy-duty plastic sheets selling in discount stores for under $2.00, water-resistant canvas ($3.75 for a three- by seven-foot cloth), or nylon ($5.50 for a three and a half- by seven-foot sheet). You can even purchase a huge seven- by twelve-foot nylon ground cloth with grommeted eyelets for under $18.00.

Weights vary: in the three- by seven-foot size, plastic weighs under a pound, canvas tips the scales at two and a half pounds, and nylon comes in at one pound even.

You can also put your own eyelets in the canvas and nylon by using a grommet kit. Eyelets are used to stake

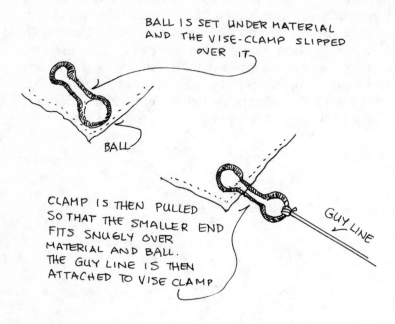

BALL IS SET UNDER MATERIAL AND THE VISE-CLAMP SLIPPED OVER IT.

BALL

CLAMP IS THEN PULLED SO THAT THE SMALLER END FITS SNUGLY OVER MATERIAL AND BALL. THE GUY LINE IS THEN ATTACHED TO VISE CLAMP

GUY LINE

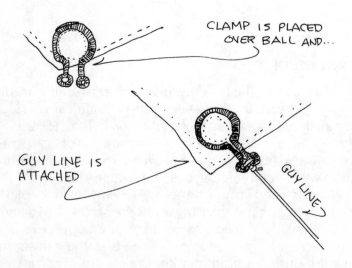

CLAMP IS PLACED OVER BALL AND...

GUY LINE IS ATTACHED

GUY LINE

Figure 5R

the ground cloth, a technique some experienced campers use to keep it tight and smooth.

Finally, the big advantage of the cheaper plastic is that it can be thrown away and easily replaced when torn or damaged.

FROST LINERS

A frost liner is a must for the year-round camper, and, if you're planning on more than just a couple of weeks in the summer, you should think of adding one to your gear. However, you then must think of buying an alpine or winter tent that has the built-in snaps for the frost liner, which is cut to size and fits specific models.

But should you decide not to purchase an alpine tent and later change your mind to expand your camping time, you can rig your own by taking a frost liner, adding grommeted eyelets, and tying it to interior poles or D-rings, which you can sew to the reinforced part of your tent.

Frost liners, depending on size, usually weigh only a pound or so. They cost from $30 to $75, but are worth the money if you plan on winter or high-altitude camping.

INSECT PROTECTION

All good tents come fully equipped with nylon mesh doorways, windows, and vent coverings, but eventually these tear and have to be replaced, especially if small children are part of the camping family. There are various grades of mesh, but the best all-around type is called "No-See-Um" mesh. This is nylon screening with holes so tiny that it keeps out the no-see-ums, tiny black flies that most people never see and realize that these insects are in the area only by the large, itchy, red welts the fly bites cause.

Mesh also comes in various colors, ranging from white to charcoal gray and black. Though the color has nothing to do with the protective aspects of mesh, charcoal gray or black are recommended, since they don't reflect sunlight as much as the lighter colors do, making it easier to look outside on a bright day. This is especially important if small children have to be constantly checked on.

The cost of mesh replacement runs from $.80 per running yard to $1.10 per running yard, but one night in the woods with no-see-ums dive-bombing your head would make $100 a yard worth the money.

POLES

Replacements for damaged or lost pole segments, A-frame connectors, cross-ridge bars, and other framework paraphernalia are available from many sources, but you should buy your replacements from the manufacturer of your tent, as these tend to fit better. Since the poles and their various accessories are so varied, prices and weights won't be quoted here. Besides, these bits of information are found in every tent catalog. Instead, a few hints on buying them will be offered here, should you need to get a replacement part sooner than you can obtain it from the manufacturer.

To begin with, if your pole is fiberglass, look for splinters or cracks, particularly at the ends. This suggests a weak pole, one that may crack in a gusty wind. With metal poles, check for burrs or sharp edges that can cut tent fabrics and guy lines. Inspect the section where the poles connect very carefully. If the metal isn't smooth, if it is unevenly finished, the pole may jam upon assembly or be hard to disconnect when packing up. Also, double-check the connections to be sure that they fit tightly with no wobble but also come apart without a hassle.

Once again, try to start with the manufacturer who

built your tent. The time or money you save by buying other brands may not be worth it when the tent collapses or pole integrity is threatened by an inferior replacement.

REPAIR KITS

Eventually fabrics will rip and other parts of your tent will be damaged in some way and you're going to have to do some repairs. Fortunately, there are many brand-name products that cover just about every fix-it job that will pop up.

Nylon Tape

Waterproof, 1.5-oz. ripstop nylon tape for tears and abrasions in nylon tents that must be fixed immediately is a necessary part of your outfit. Usually pressure sensitive, it comes in various colors and the price averages out to around seventy-six cents per five-foot roll. Nylon tape suitable for all kinds of fabrics is also available for approximately fifty cents per running yard.

Canvas Repair Kits

Complete with small patches of canvas, thread, sewing needle, and small bottle of canvas cement for about $1.50.

Fabric Cement

Perfect for fixing canvas tears or abrasions without sewing. A half-point bottle with brush-on cap costs about $3.

Grommet Kits

For making or replacing damaged grommets; includes different-sized grommets, hole cutter, hardwood block, inserting punch, and die. The grommet sizes run from ¼ inch to ½ inch and all the standard sizes in between. Price range is $4.20 to $7.

Sewing Awl Kit

For repairing floors and reinforcement patches and adding features like D-rings, loops, or inside pockets that require stitching. Cost: $4.

Snap Fastener Kits

For replacing or installing snap fasteners on the tent fabric. Cost: $1.25 to $1.75.

Waterproofing

Should you want to waterproof your tent or repair a leaky rain fly, there are chemicals that can be sprayed or brushed on, depending on how much you're going to use. The brush-on type goes for around $4 a gallon, or $2 a quart, and the handy spray cans sell for about $3. There are dozens of brand names to choose from, and your camping supply store has plenty on the shelves.

Seam Sealants

A little more than $2 for a three-ounce tube. A must to pack along, especially if rain is expected.

Loops/D-Rings

These can be slipped over the ridge pole, if your tent

has one, or sewn to various parts of the tent. Handy for hanging up wet clothing and small gear. They run from thirty-five cents each for fabric loops to sixty-five cents for rust-proof D-rings.

***

Other accessories that should be purchased are whisk brooms (used for cleaning off tents before packing; this prevents tears and damage due to forest debris caught in the strands of the fabric), zippers, Velcro replacement patches and strips, snap-lock replacements, and just about any other item needed to improve the quality of your tent. Check with your camping supplier and ask those who have been out many times what's best for repair, replacement, and additions.

6

Tent Classifications

Obviously, if tents weren't so technically designed, they would come in one basic model and the only factor that would concern the buyer would be how large a tent was necessary. Then the buyer could merely compute the size of the shelter needed to house the number of people in the party and buy accordingly. But tents have become complex and varied, so selecting a tent is not so simple. To help the buyer, tent authorities and other outdoor experts have classified tents into categories that help explain their basic design, function, and features. The reader should realize that the borders between these groupings are blurry and, say, a forest tent can be used as an alpine model and vice versa. However, this classification helps the buyer determine his needs, because, although the tent styles may overlap, each category offers specific features for particular needs and this knowledge will aid in any decision making.

But before the groupings are discussed, a few features that all quality tents offer will be examined. Knowing these features will help the buyer weed out the almost-good-enoughs from the true quality shelters. Some of these items, like breathability, natural water repellency, tub floors, and rain flies, have already been covered, but one very important aspect, ventilation, has been saved until now.

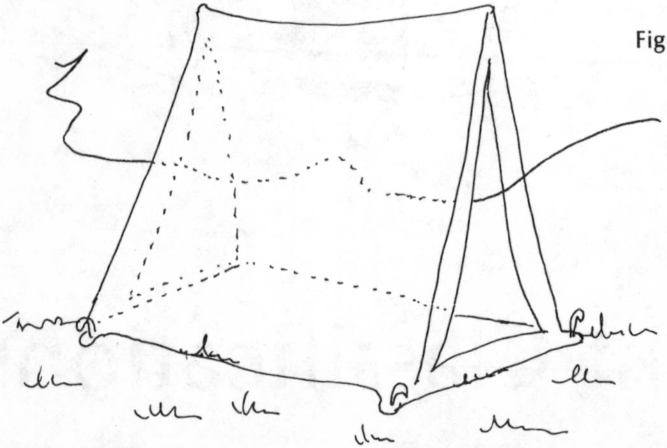

Figure 6B

VENTILATION

This often overlooked aspect of tent design is the one that can make the difference between a comfortable night or a miserable one, even in the summer. And in the winter, ventilation can mean the difference between life and death in some instances.

Vents are employed in addition to using a breathable fabric. At least two vents are needed, one at each end of the structure, for proper cross ventilation (see Figure 6B). In some models, the front entrance (or just the top of the doorway, if the weather is rainy or cold) can be used as a vent, and the rear vent is a comparatively large window (or, again, just its peak); in some types, the rear vent is just that, a circular hole in the back of the structure (see Figure 6C). Either type of rear vent and the front entrance, naturally, should be of the double-slider zipper variety and should be covered by nylon mesh. This way, the tip of the entrance and window (or vent) can be partly opened, no matter how harsh the weather.

Vents should be placed high up, just under the ridge line, so that they can remain open in wet weather, protected by the rain fly. Positioning these vents (called ridge vents, logically) high up also allows the stale, warm air that rises inside the shelter to escape. Some manufacturers install what they call chimney vents lower down on the structure so that the cooler outside air can be drawn in and push out the stale air. The premise is sound, but in rainy weather the lower vents have to be closed to keep out the rain, and this cancels out the benefits chimney vents provide.

Cooking in a tent will add moisture as well as dangerous fumes such as carbon monoxide to the inside atmosphere, so an exhaust vent located directly over the cooking area is a must (see Figure 6D). Some models feature tunnel-shaped exhaust vents that can be pulled

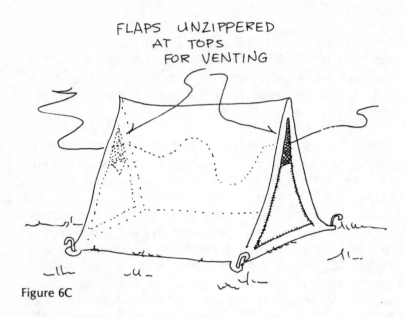

Figure 6C

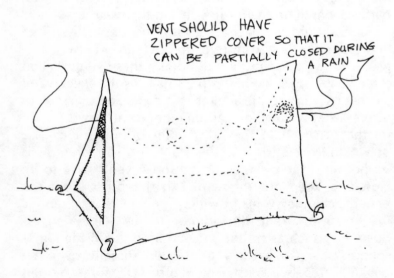

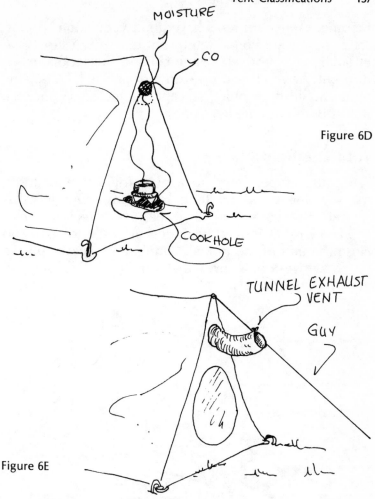

MOISTURE

CO

Figure 6D

COOKHOLE

TUNNEL EXHAUST VENT

GUY

Figure 6E

out and moved around to prevent wind from blowing inside (see Figure 6E). Cooking inside a tent will throw a lot of moisture into the air, not all of which will escape unless that exhaust vent is at the right spot, and you might end up with frost covering everything the next morning if the carbon monoxide from the cookstove doesn't kill you first.

Some waterproof tents have good ventilation systems, so if you are sold on a waterproof tent, make sure it has enough vents, both ridge and chimney, to let that moisture out, or you'll end up with a drippy tent. All-plastic tents should be avoided, except as emergency shelters, unless both ends are kept open.

TUNNEL ENTRANCES

Because zippers tend to jam in frigid temperatures, many alpine or winter tents have tunnel entrances. A tunnel entrance is generally three to four feet in diameter. Drawstrings instead of zippers are used to open and close the end of the tunnel. Tunnel entrances also have four- to five-foot sleeves that can be pulled out and guyed (see Figure 6F).

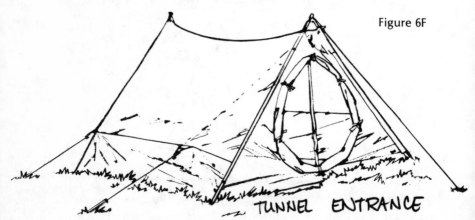

Figure 6F

TUNNEL ENTRANCE

COURTESY EUREKA!

The benefits of a tunnel entrance are that it doesn't freeze or jam shut; there's less chance of snow drifting to prevent entry or exiting than with the vertical entrance; it's an escape route, should drifting snow prevent using the regular entrance; you can enter the tent without letting in as much cold air or wet snow as you would with the full-sized entries; and it can be linked to another tunnel entrance, making a two-room shelter that can come in handy on days when you're snowbound for a long period of time (see Figure 6G). The big disadvantage is that you have to crawl in and out.

Figure 6G

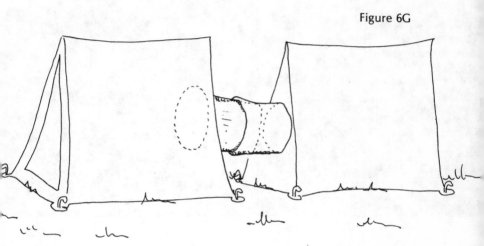

VESTIBULES

Some tents come with one or both ends designed so that they aren't flat-sided, or vertical, but can be pulled out to form vestibules for extra room (see Figure 6H). Vestibules can be full canopies or split canopies, and some have tunnel entrances in them. Vestibules can come with a tub floor like the living quarters of the tent, but if they don't, there should be a fabric sill between them and the rest of the tent to keep forest debris, mud, or snow out of the sleeping area. Bare-floor vestibules can also be used as cooking areas if they are vented, and both versions provide extra storage space for gear and wet boots.

Figure 6H

COOKHOLES

During inclement weather, people often cook inside their tents. If this is a possibility in your plans, you should consider a tent with a cookhole, a zippered semicircle in the waterproof floor that allows you to cook right on the bare ground. This feature allows for spillage without complications. If you're cooking on a regular tub floor and should knock over the pot, the food will ooze around and eventually be sopped up by your pack or your sleeping bag. You can also damage the nylon floor with heat from the stove. But on bare gound, the heat and any spillage are absorbed by the earth.

If you use a ground cloth, you can still enjoy the benefits of a cookhole by cutting a hole in the ground cloth. The small opening won't cancel out the other advantages of a ground cloth.

However, don't think of a cookhole unless there's an exhaust vent above it. The convenience isn't worth the risk.

Finally, the cookhole, especially in a blizzard, can double as a garbage disposal or even an emergency bathroom. Merely dig a hole in the snow, throw your wastes in it, and cover it back up, zipping shut the cookhole.

Although opinions may vary, this ends the list of basic necessities when it comes to quality tents. If you are intrigued by other features not detailed in this book, manufacturers' catalogs or the advice of experienced campers will provide the data you may need in making a selection.

TENT CLASSIFICATIONS

There are seven basic classes that tents fit into. They are: emergency; forest, or woodland; alpine, or winter (also known as tundra, mountaineering, or timberline); wind; rain; desert; and tropical. These designs can be found in the family or backpacker sizes. As already mentioned, with modifications, they can be used in different environments and terrains.

EMERGENCY

Also called instant tents, inflatable tents, or tubes, these are waterproof, all-plastic tents. The waterproof tarps also known as emergency shelters will be covered in detail in Chapter 8, *Other Shelters*. This section will be concerned only with the plastic that's sewn or cast to form a typical tent structure.

The instant and inflatable-frame tents are the ones that

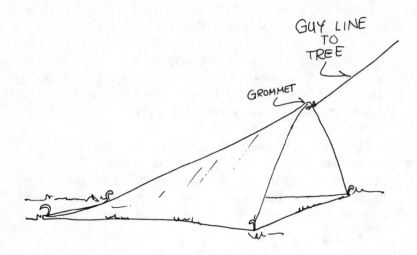

Figure 6l

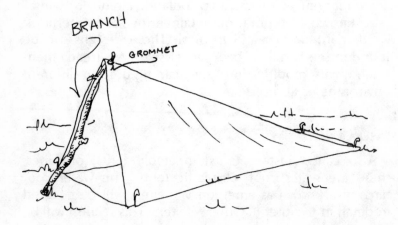

are just shaken out and erected with cord or deadwood or plastic supports that are inflated by mouth (see Figure 6I; also see Figure 3DDD in the chapter on frames) and staked. They are constructed out of heavy-duty waterproof plastic and have a waterproof floor. Some have flaps that can be closed, but most just have an open front end, since they are designed primarily for emergency use. The average size is 6'6" long by 3' wide. When erected, they are 34" to 40" high at the open end and taper down to a few inches at the other end. Costing about $3 each, the instant tents can be folded or rolled into a pocket-sized package of only 7" × 3", and the total weight is 14 to 15 ounces.

Tube tents (see Figure 6J) are exactly what the name implies. Though they are heavier (slightly more than a pound) and cost more (around $5) than the instant tents, they are made of stronger, thicker plastic and are roomier. They also have flaps that can be *partly* closed to keep out most of the rain, but it should be noted again that you should never completely seal off both ends of a plastic tent. It can't breathe and neither will you for long if you close it up tightly. Tube tents, like the instant ones, are cheap enough to be disposed of after a couple of uses. They should be thought of as emergency shelters only, since they don't breathe, don't have nylon mesh to keep out the bugs, tear easily, and are so unstable that a high wind can rip the tent from its moorings.

FOREST, OR WOODLAND

This multipurpose tent is the most popular with campers because of the variety of environments it can be adapted to; still, it is called a "forest tent" because most tentmakers designed it—and most tent buyers utilize it—for forests, woodlands, and sheltered grassy areas in mild weather. Consequently, although these tents can be

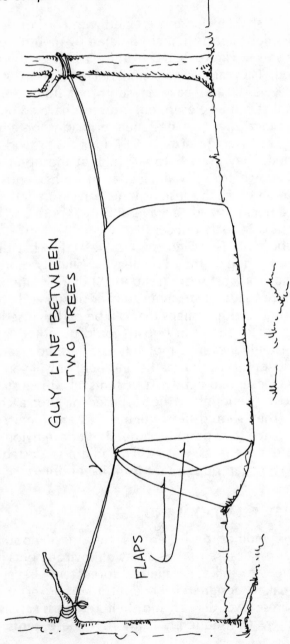

GUY LINE BETWEEN TWO TREES

FLAPS

Figure 6J

modified to fit into other classifications, the details that follow are set up on the basis of typical usage in forests or woodlands. This tent is aimed at the summer-only camper who seldom meets Mother Nature at her worst.

Size. These tents come in both family and backpacker sizes. Since most of the camper's time is spent outside the shelter, it is used mostly for sleeping and protection against short periods of inclement weather and, therefore, the forest or woodland tent can be a little smaller without sacrificing comfort. Incidentally, if you find sizes expressed in feet and inches confusing, all tentmakers tag their structures as "one-man," "two-man," "six-man," and so on. This indicates how many average-sized adult males can fit into the structure without piling them up like fireplace logs. If there are, for example, two adults and two small children in your family, a four-man tent should be large enough. You could probably even get away with a three-man structure, although you had better hope that the kids are so exhausted at night that they won't roll all over you.

Frames. The frames available for backpacker tents are I-pole, A-frame/I-pole combination, and double-A-frame with or without ridge pole; for the larger tents there are center A-frame or I-pole, box, and umbrella. Some exoframes, like the Draw-Tite, come in a size range from one-man to six-man models. Finally, since most of these frames form the wedge-shaped tent with a sloping canopy, forest tents readily shed rain if they are properly rigged with rain flies.

Fabrics. Taffeta, ripstop nylon, or pima cotton is recommended for the backpacker, although the heavier cotton canvas can be used. Either canvas or nylon is recommended for family size tents, depending on your personal preferences. Both sizes come in waterproof materials and Gore-Tex as well as the more popular double-wall construction designs. Some tents have a mesh nylon for a canopy to be used without a rain fly for summer nights and with a fly for the rainy ones. However, this design doesn't allow for dew, and you may find your equipment damp from this natural condensation if a fly isn't erected. Also, tents with mesh canopies cannot be converted to spring, fall, or winter usage. All tents should have tub floors unless you're going into the sidewall or Baker tent, and then a ground cloth should be employed.

Wind Stability.   Forest tents are surprisingly good in this respect, but wind isn't generally much of a factor, since the typical usage is in areas where trees, bushes, brush, hills, and valleys provide natural windbreaks.

Necessary Accessories.   To save yourself the maddening effects of clouds of carnivorous insects, mesh should cover all openings, whether they are entrances, windows, or vents. Since the ground is usually soft where forest tents are used, lightweight stakes like plastic, aluminum half-tubes, "U" staples, or skewers can be used. Fabric tighteners may be needed, but this is doubtful, since wind and rain aren't deciding factors. Besides, a well-designed tent won't need extra help in keeping it taut in mild weather. A ground cloth can be employed, but this depends on your preference.

ALPINE, OR WINTER

These are the tents that are created for the harsher climates and weather conditions. Since being stuck inside the structure is common during winter or mountain side trips, some models, like the dome, have more room inside than forest tents with the same floor space. Make sure that any alpine tent you buy has a good ventilation system.

Size.   These are produced in both the backpacker and family sizes, but most winter campers prefer the smaller tents, especially those with tunnel entrances that can be connected together.

Frames.   Usually the double A-frame (with or without ridge pole), Draw-Tite, dome, and tunnel frames are recommended, as these frames are self-supporting as well as having the ability to shed rain and withstand wind. Self-supporting frames are important, since the ground may be too rocky, too frozen, or too covered with snow for usual staking, and deadman anchors may have to be employed.

Fabrics.   The fabric used in winter tents is the same as that used in forest tents, with the addition of a rain fly that extends almost to the ground. This is necessary to avoid seepage from heavy rains or melting snow. Gore-Tex is fast becoming a favorite among tent buyers who prefer single-wall construction. Snow or sod flaps should be always considered.

Wind Stability.   This is one of the most important factors to be considered. The dome, Draw-Tite, and tunnel models fit the specifications best.

Necessary Accessories.   As surprising as it may sound, "No-See-Um" nylon mesh is a must if you're going to be traveling in the north country during the spring or fall. Hordes of mosquitoes, black flies, and no-see-ums have been known to drive herds of thick-skinned buffalo mad in these areas, so if you don't want to end up running off some cliff with a wild look in your eye, make sure a top-quality nylon mesh covers all tent openings. Cookholes, always in conjunction with exhaust vents, are features to chalk in on your need-board, as you might have to cook inside because of the weather. Tunnel entrances and vestibules are good features to look for, and a frost liner should be added. If the cost of a frost liner is prohibitive now, make sure you purchase a tent with the frost liner connections, usually snaps, so that you can add a liner later without having to revamp the tent. Stakes should be steel for the rocky terrain, and the snow stake is a must. Fabric tighteners may be handy, although most alpines are very tight once erected.

WIND

As the name implies, wind tents are structures aimed at efficiently spilling the steady winds encountered on the sides of mountains and unprotected sea coasts.

Size.   The smaller two- or three-man tents are usually preferred, although some manufacturers produce a four- to five-man size.

Frames.   Dome and tunnel structures are used most of the time, although a center I-pole with ground plate is employed in the larger versions. Double A-frames are also used by some campers who swear by the more traditional frames.

Fabrics.   The covering of a wind tent can be any of the fabrics already mentioned for the other classes, with your personal preference being the deciding factor. However, some old-timers claim that the smoother nylon, both taffeta and ripstop, is better for wind tents because the air can glide more easily over the smoother surface.

Wind Stability (WS). Obviously, WS is the most important aspect in this class. A wind tent should present a low profile to the wind, whether in gusts or a steady blast. It should have extra reinforcements at important stress areas and be wrinkle-free after erection. Also, the shape should be designed so that the air can move over and around without hitting any flat surfaces. Consequently, the dome and tunnel seem to be the best designs, though the A-frame still has its admirers.

Necessary Accessories. Extra guy lines and stakes may be necessary. The latter should match whatever terrain the tent will be standing on. If the dome or tunnel is used, there's not much reason for fabric tighteners, since these structures are taut enough when rigged properly.

RAIN

Again, the name describes the classification, so ability to shed rain, breathability, good ventilation, and rain flies that can be guyed close to the ground count. Roominess is also an important factor.

Size. Since most tents shed rain effectively enough to be used in wet weather, this isn't really a separate category. When looking for a tent to use in rainy environments, consider size, as in wet weather you may be trapped in a tent for long periods of time. Family-size models provide the roominess necessary for comfort under such conditions.

Frames. All frame designs shed rain, but the wedge shapes with close-fitting rain flies are the most effective, even if wind is a factor.

Fabrics. Again, the same fabrics recommended for

other tents, as long as they breathe and a rain fly is employed, work effectively. However, it should be repeated that the rain fly should be guyed near the ground to avoid splattering water on the nonwaterproof sides of the tent above the tub floor. The fly should be rigged so that there's a space of air between the canopy and the fly to cut down on the inside condensation. Ventilation is important, since the tent may have to be sealed against the weather; all vents should be protected by the fly yet be free enough to allow constant emission of the inside atmosphere. Naturally, a good tub floor is important.

Here, someone might argue for waterproof materials, but since they don't breathe and since the tent openings will have to be closed against the weather, condensation is a bigger problem than the outside wetness. Some manufacturers claim to have solved this by ventilation systems, but the consensus of expert opinion is that the double-wall construction is still the best.

Wind Stability. This factor should be considered, but if the terrain is a sheltered one, just about any structure that spills wind fairly well will suffice.

Necessary Accessories. Unless you're planning to spend weeks in a monsoon, cookholes aren't really necessary. If you do decide to buy a tent with this feature, again, check for the exhaust vent. Stakes should be nonrusting aluminum or plastic and the kind that will hold in mushy ground. Sod flaps would be helpful, as rocks could be placed on them. Cord tighteners should be standard equipment for any camper, but if rain is expected, they are a definite must. This is because cord and fabrics tend to stretch in a steady downpour, necessitating constant adjustment of the guy lines to keep the tent and especially the fly tight and wrinkle-free so that the water doesn't have a chance to collect anywhere and instead runs off onto the ground.

DESERT

These tents have several factors to be weighed, but the most important are fabric and frame.

Size. Experienced desert campers prefer the larger sizes, since they are cooler, but desert tents can be found in both family and backpacker sizes.

Frames. Since guying and staking is difficult in sand, self-supporting frames like the dome, tunnel, and double-A are recommended. Some desert rats like the double-A with sod flaps so that they can pile rocks or shovel sand on them in the absence of guys and stakes.

Fabrics. This is probably the most important consideration of all. Wind is common in the desert, so a top-quality, tightly woven fabric is a must to keep out the wind-driven sand. Rain flies are also employed, both to shield the tent from the sun and to provide a space of dead air to insulate the tent, keeping it cooler in the daytime and warmer at night.

Wind Stability. As already mentioned, designs that allow wind to glide around the structure are best. Sod flaps add extra stability.

Necessary Accessories. Sand (snow) stakes are important in desert camping where they can be used, and even deadman anchors may be needed. Cookholes may be desired, since this solves the problem of bits of grit that always land in food cooked outside as well as providing an inside hole to sweep out sand. Needless to say, good venting, especially exhaust vents, is important. Though there aren't too many mosquitoes on the desert, there are flies and fleas, so nylon mesh should cover all openings.

TROPICAL

Here, the nylon mesh canopy mentioned in the section on forest tents may be ideal, as long as all openings are completely protected from bugs. This also includes the floor and the bottom of the entranceways.

Size. Both family and backpacker versions are manufactured in this class, and no particular size is suggested. Here, size depends on your needs.

Frames. Self-supporting frames are helpful but not necessary, even though the ground may be too soft to hold the tension of a guy line.

Fabrics. The fabric must have breathability, to avoid inside condensation. A rain fly is also important, as the outside dew is heavy in hot, humid areas.

Wind Stability. This is not an important factor, since most tropical terrains have plenty of natural windbreaks, but the self-supporting tents that are recommended for this environment also effectively spill wind if any should spring up.

Necessary Accessories. Rustproof stakes and a ground cloth are suggested. Be sure there's plenty of nylon mesh covering all openings and that the bottoms of the entranceways are also protected from the multitude of crawling bugs.

Once again, the reader is reminded that tent classifications tend to overlap and that with modifications, a tent listed in one class can be employed in a different environment. The classifications are merely based on typical usage and are intended to give the reader a clue as to what kind of tent is displayed on the sales floor.

7

What's on the Market?

This chapter will concentrate on what is available to today's tent buyer. Photographs and manufacturers' specifications on weight, types of frames and fabrics, and so on will illustrate what some of the tentmakers have to offer in both the large, family-sized structures and in the category aimed at the backpacker.

However, it must be emphasized that the publication of the following pictures and data is *not* a recommendation. This was not the intention of *The Complete Tent Book.* Instead, the information is meant to depict the various styles and designs currently on the market and give the buyer some idea of what his dollar will buy. It is geared to be nothing else and certainly not a recommendation of one type of tent or tentmaker over another. In fact, the companies are listed alphabetically to avoid any hint of favoritism.

Also, the data do not include all the tents marketed by a particular manufacturer. Due to the limitations of space (and the fact that you can write away for a company's catalogs by using the addresses at the beginning of each section), only a representative sampling is covered, providing examples of designs that should satisfy the needs and wants of the majority of tent buyers.

If you encounter a tent or manufacturer not listed in this book, that doesn't necessarily mean the tent isn't a quality product or the company isn't a reliable firm. Unfortunately, some tentmakers didn't reply to the request for information on their tents; others answered too late to make press time; and a few were not contacted because they had just started in business, because their tents were similar to others, or because they were more of a supplying company than a manufacturing and designing one. Since *The Complete Tent Book* has detailed all the features that determine a quality shelter, you should be able to compare and evaluate a structure not listed in the next two sections. A good tent is usually made up of the same components, no matter who puts them together. Finally, to be completely fair, a listing of manufacturers who responded a little too late will end this chapter.

Another point to be recognized is that most tents described here fall into the categories of forest, alpine, or wind. If you're interested in tents in other classifications, write to the manufacturers listed here and ask if they produce the frame, fabric, or other necessary accessories that make up that special type of tent. They'll be more than happy to answer any request.

Finally, one last hint on determining size needed: if you're going to be a backpacker, stick to the old axiom that says never to carry more than three pounds of tent per person. If there are only two of you hiking through rough country, stay with the tents that are six or seven pounds or less; three in your party means that the shelter should be no heavier than nine or ten pounds; and so on. With the larger, family-sized shelters, the kind of tent that will be hauled to the site by vehicles, weight doesn't matter very much.

***

The following specifications were supplied by the manufacturers. The method of listing the various items is as follows:

1. Size is listed first by the number of adult males the tent will house and then in feet and inches, measured around the perimeter of the tent floor. The ridge height will be included if that information was supplied.

2. All tents can be assumed to have fine nylon mesh covering all openings and nylon zippers unless otherwise noted.

3. All fabrics are fire resistant unless noted, and all tents have a waterproof boat, or tub, floor unless otherwise stipulated.

4. The price range is approximate, as tents may tend to cost a few dollars more in different parts of the country. The prices listed don't include local or federal taxes.

Brigade Quartermasters, Ltd.
P.O. Box 108
Powder Springs, Georgia 30073
(404) 943-9336

Pak-Kot (first view) (Brigade Quartermasters)

PAK-KOT

Size
One-man: 75" × 25", with a 7" ridge line.

Frame
Aluminum with steel spring legs on the cot.

Fabric
Cot: nylon; canopy: Gore-Tex; sidewalls: treated, waterproof ripstop nylon. No floor.

Features
An unusual and radical design in a one-man shelter that is lightweight and comes with a stuff sack for all the parts.

Weight
Nine lb., 4 oz.

Price
$130 complete.

Pak-kot (second view) (Brigade Quartermasters)

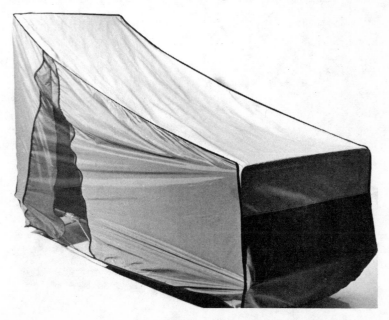

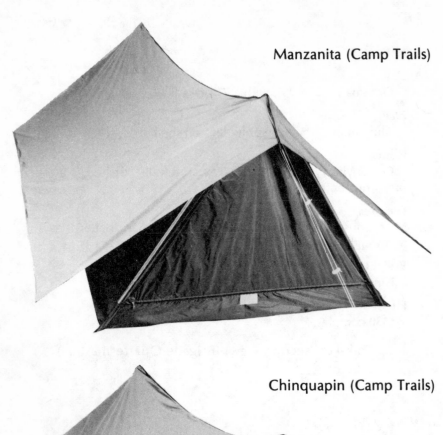

Manzanita (Camp Trails)

Chinquapin (Camp Trails)

Camp Trails Company
4111 W. Clarendon Avenue
Phoenix, Arizona 85019
(602) 272-9401

MANZANITA

Size
Two-man, 7' × 4'8", with a ridge height of 3'6".

Frame
Shock-corded aluminum A-frame in front and I-pole in rear.

Fabric
Canopy is nylon mesh and sidewalls are nylon. Fly is waterproof nylon.

Features
Good for warm nights, providing plenty of insect protection. Colors: tent—brown; rain fly—red, orange, or green.

Weight
Five pounds.

Price
$114

CHINQUAPIN

Size
Two-man 7' × 4'8", with a 3'6" ridge.

Frame
Shock-corded aluminum A-frame in front and I-pole in rear.

Fabric
Canopy and sidewalls are ripstop nylon.

Features
Small screened ridge-line window for ventilation; full rain fly; tub floor extends to 8" up the side to avoid splash. Colors: tent—brown or tan; rain fly—red, orange, or green.

Weight
Five lb., 4 oz.

Price
$130

***

Camp Trails also produces a 7 lb., 7 oz., three-man tent called the Sahuaro that sells for $185.

Classic (Coleman Company)

The Coleman Company
Wichita, Kansas 67201
(316) 261-3211

CLASSIC

Size

Five-man is 11' × 10' with ridge height of 7'; four-man is 10' × 8' with ridge of 6'6"; three-man is 8' × 6'9" with ridge of 5'; two-man backpacker is 7'9" × 5' with ridge of 42".

Frame

Color-coded aluminum box with ridge and eave poles; self-standing (although it is suggested that all self-supporting tents be staked anyway for true tent security whenever possible).

Fabric

Canopy: 7-oz. drill; sidewalls: spun polyester; tub floor with rope stake line in larger models; two-man is coated ripstop nylon.

Features

Five-man: double front door and 2 triangular side windows (61" × 39" each); four-man: double front door and 2 triangular windows (48" × 44" each); three-man: double front door with one rectangular rear window (24" × 41"); two-man: zippered side door (34" × 37") with semicircular ridge-line top vent with weather covering. Colors: canopy—white; side walls—tan; floor—green.

Weight

Five-man: 36 lb.; four-man: 35 lb; three-man: 19 lb.; two-man: 5 lb.

Price

Available upon request.

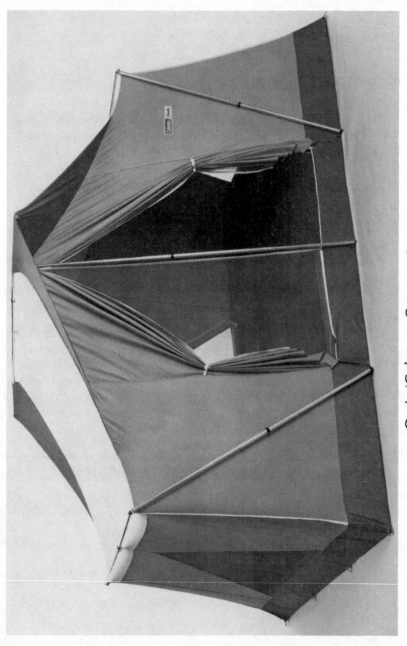

Oasis (Coleman Company)

OASIS

Size

Six-man: 13' × 9' with ridge of 7'6"; five-man: 12' × 8' with ridge of 7'6".

Frame

Color-coded aluminum box with ridge, center, and eave poles; self-supporting.

Fabric

Canopy: 7-oz. drill; sidewalls: 7-oz. combination of 80 percent cotton and 20 percent polyester.

Features

Six-man: double front door with 3 triangular windows (53" × 40" each); five-man: double front door with 3 triangular windows (46" × 40" each). Colors: canopy—white; sidewalls—green; floor—green.

Weight

Six-man: 52 lb.; five-man: 48 lb.

Price

Available upon request.

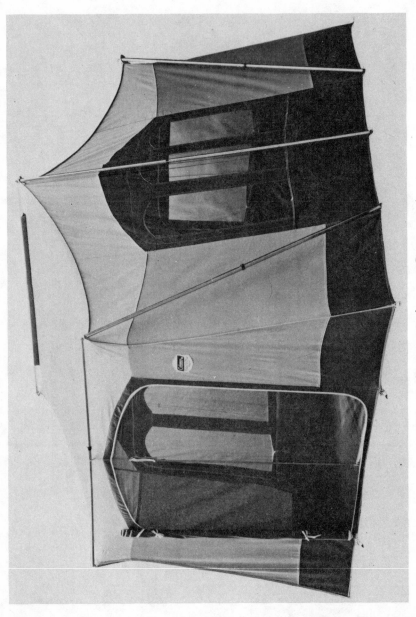

Villa Del Mar (Coleman Company)

VILLA DEL MAR

Size
12'× 9', with ridge of 8'.

Frame
Color-coded aluminum box with ridge and eave poles; self-supporting.

Fabric
Canopy: 7-oz. drill; sidewalls: 7-oz. coated drill.

Features
Flat, zippered, weather-tight door; one 27" × 44" window and six 16" × 39" windows. Colors: canopy—cream; sidewalls—gold and khaki; floor—brown.

Weight
Sixty-three lb.

Price
Available upon request.

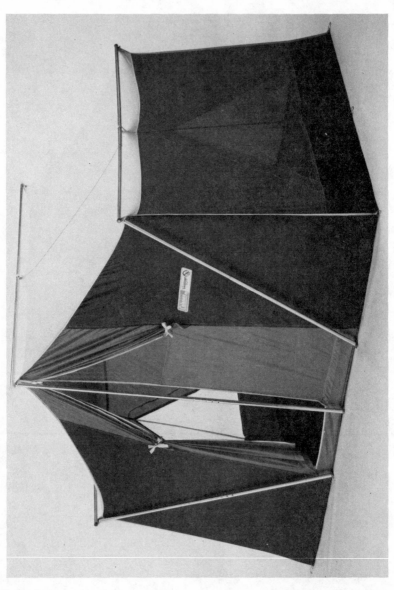

Vacationer (Coleman Company)

VACATIONER

Size

Five-man: 11'2" × 8'2", with a ridge of 6'6"; four-man: 9'8" × 7'2", with ridge of 6'6".

Frame

Color-coded aluminum box with ridge, center, and eave poles; self-supporting.

Fabric

Canopy: 6.5-oz. drill; sidewalls: ripstop nylon.

Features

Five-man: double dutch front door and 3 triangular windows (53" × 30" each); four-man: double dutch front door and 2 triangular windows (47" × 30" each). Colors: canopy—white; sidewalls—red; floor—green.

Weight

Five-man: 29 lb.; four-man: 26 lb.

Price

Available upon request.

Terrace (Eureka)

Eurekal Tent, Inc.
625 Conklin Road   (607) 723-7546
P.O. Box 966
Binghamton, NY 13902

TERRACE

Size

Five-man plus: 10' × 10', with a center ridge height of 8', front ridge of 6', and a rear ridge measurement of 5'. Awning is 10' × 7'.

Frame

Aluminum box with ridge and eave poles as well as awning poles; semi-self-supporting.

Fabric

Canopy, sidewalls, and awning are 7-oz. poplin.

Features

The awning is attached, thus doing away with having to erect it separately; the waterproof floor extends 8" up the sides; a full-sized awning net is available; comes complete with tent, poles, pole bags, stakes, and guy lines; 3 large triangular windows and double front doors provide excellent ventilation. Color: pearl gray.

Weight

Fifty-six lb.

Price

$265.50 for the NFR fabric; $289.95 for the FR; awning net enclosure is $47.50.

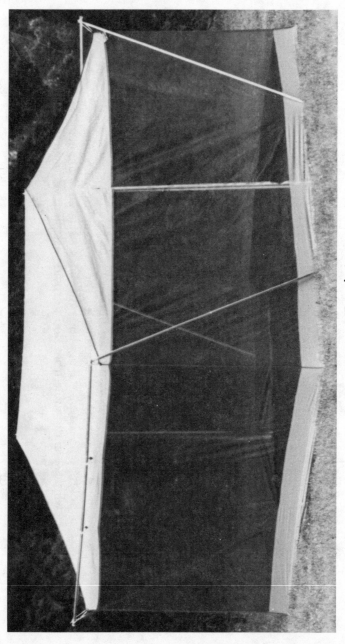

Screen House (Eureka)

SCREEN HOUSE

Size

SH 1210 is 10′ × 12′; SH 1410 is 10′ × 14′. Both sizes have center ridges of 7′8″ and wall heights of 5′8″.

Frame

Aluminum box with interior ridge pole and exterior eave poles.

Fabric

Canopy is made of 7-oz. poplin or 7-oz. nylon. The all-nylon mesh sidewalls are heavy-duty, web-reinforced on the corners (stress points), and seamless.

Features

The screen house can be used as a tent or an enclosed dining area in the backyard. It is large enough to hold two picnic tables and if used in conjunction with a tent, the large openings can be attached so that you can walk from the tent right into the screen house. Outside webbing allows you to attach a nylon curtain to keep out the rain and wind. The curtain comes in three sizes to fit all or part. Color: tan.

Weight

SH 1210: 32 lb.; SH 1410: 35 lb.

Price

SH 1210: $146.95 for NFR, $155.50 for FR cotton, and $132.50 for FR nylon; SH 1410: $174.50 for FR poplin and $144.50 for FR nylon. Wind curtains range from $12.95 to $14.95.

Great Western (Eureka)

GREAT WESTERN

Size

Comes in three sizes, known as the 8, the 10, and the 12. The 8 is 7'8" × 7'8" with a center height of 78", 66" at the eaves; the awning projects out 66". The 10 is 9'6" × 9'6", with a ridge height of 84" at the center and 72" at the eaves; the awning projection is 72". The 12 is 117" × 117" with a ridge height of 76" at the eaves and 92" in the center; the awning projects out 78".

Frame

Shock-corded aluminum umbrella with I-poles for the awning; self-standing.

Fabric

Canopy: 1.9-oz. ripstop nylon; sidewalls and fly are waterproofed 1.9-oz. ripstop, as is the awning.

Features

Double-wall construction; three large windows, a walk-through door, and a roof or ridge vent provide excellent ventilation; extra reinforcements at stress points; fabric hung from frame by reinforced shock cords; four corner rings and a lantern hook are provided on the inside. Color: willow green/ yellow.

Weight

The 8 is 18 lb., 3 oz.; the 10 is 25 lb., 1 oz.; the 12 is 27 lb., 4 oz.

Price

Eight: $189.95; 10: $225; 12: $275.

Space Tent (Eureka)

SPACE TENT

Size

Two sizes are offered in this model, the Space 10 and the Space 12. The 10 is 10' × 10', with a ridge height of 8' and an eave height of 6'. The Space 12 is 11'9" × 11'9" and is also 8' at the center ridge and 6' at the eave ridge. The awnings for the 10 are 7' × 10'2"; for the 12, 8' × 12'.

Frame

Shock-corded aluminum umbrella (exterior), with I-poles for awning.

Fabric

Entire tent is constructed of 7-oz. poplin with a water-and-mildew-resistant treatment that adds no weight to the material. The tub floor extends 14" up the side of the tent.

Features

Awning can be covered with a net enclosure or a canvas curtain; large double door and three large triangular windows plus a ridge vent provide excellent ventilation. Color—tan, with a green floor.

Weight

Space 10: 54 lb.; 12: 65 lb.

Price

Space 10: $244.95 for NFR fabric, $262.50 for FR material; Space 12: $274.95 for NFR, $299.95 for FR. Mesh enclosures: 10: $39.95; 12: $45.95. Canvas curtains: 10: $74.50; 12: $92.95.

Highlite (Eureka)

HIGHLITE

Size

Two-man: 5' × 7' with a 44" front height and a 33" rear height.

Frame

Shock-corded aluminum A-frame in front and I-pole in rear; requires guying and staking for extra tent security.

Fabric

Canopy is 1.9-oz. breathable ripstop; the lower sidewalls, floor, and fly are 1.9-oz. waterproof ripstop.

Features

A lightweight backpacker tent; floorless vestibule zips into half-vestibules; ventilation is provided by large rear window and front entrance. Colors: orange and blue or green and yellow.

Weight

Six lb., 3 oz.

Price

$89.95

Aleutian/Yukon (Eureka)

ALEUTIAN/YUKON

Size

Aleutian: 86" × 96", with a 52" height; Yukon: 60" × 90", with a 48" height.

Frame

Fiberglass dome; completely self-supporting, but tent and rain fly should be guyed in extreme weather conditions.

Fabric

Canopy is 1.9-oz. breathable ripstop nylon; 1.9-oz. waterproof floor and fly.

Features

Igloo entrance allows entry and exit without letting the outside weather in; rain flies are connected by shock cords; the Aleutian has two "A" doors, providing two entries; the Yukon has one "A" door and a large rear window that can be opened either at the top or bottom for ventilation. Color—willow green and yellow.

Weight

Aleutian: 9 lb., 12 oz.; Yukon: 7 lb., 9 oz.

Price

Aleutian: $189.95; Yukon: $139.95.

Back Country (Eureka)

BACK COUNTRY

Size

Two-man, with a length of 96" and a front and back width of 42"; height at peak: 64"; front 54"; rear: 12".

Frame

Exterior center A-frame with staking and guys to draw out material.

Fabric

Canopy: 1.9-oz. ripstop (breathable); floor and fly: 1.9-oz. waterproof ripstop.

Features

Double-wall construction; ventilation provided by single front door and ridge vent with storm cover. Color—orange and blue or green and yellow.

Weight

Five lb., 12 oz.

Price

$108.50

Nu-Lite (Eureka)

NU-LITE

Size
Two-man: 5' × 74", with a ridge height of 42".

Frame
Sectional double I-poles, front and rear.

Fabric
Water-resistant taffeta nylon.

Features
One of the most economical two-man tents produced; side pullouts provide extra room inside; large door and ridge vent for ventilation; comes in assorted colors.

Weight
Three lb., 12 oz.

Price
$43.50

Expedition Timberline (Eureka)

EXPEDITION TIMBERLINE

Size

Two-man: 5'3" × 7'2", with a ridge height of 42".

Frame

Double A-frame with ridge pole, all aluminum.

Fabric

Canopy: 1.9-oz. ripstop; lower sidewalls, fly, and floor are 1.9-oz. waterproof ripstop.

Features

Optional snow or sod flaps; front vestibule with tunnel entrance; tunnel exhaust vent over optional cookhole; zipped "A" doors; interior clothes rings and map pocket; reinforced with double-strength nylon at all stress points; fly attaches to frame with brass "S" hooks and shock cords for added stability; optional frost liner. Color: caramel and sandstone.

Weight

Eight lb., 14 oz.

Price

$189.95; frost liner: $37.50; snow flaps: $16.95; cookhole: $12.50.

Draw-Tite (Eureka)

DRAW-TITE

Size

One of the largest ranges of sizes in one design. One-man: 42" × 36", with a 38" front and 27" rear; two-man: 5' × 7'9", with a 48" front and 30" rear; three-man: 7' × 8'10", with a 5'4" front and a 3'2" rear; four-man: 8' × 10', with a 6'6" front and 50" rear; six-man: 9' × 12'3", with a 6'6" front and 50" rear. All are 7 feet long.

Frame

The uniquely designed Blanchard type of aluminum exterior frame does away with guys, poles, and stakes (although some experts claim guying and staking in extreme weather is always advisable); completely self-supporting.

Fabric

Canopy is 7-oz. poplin with a 1.9-oz. waterproof nylon floor and fly.

Features

Exterior framework means interior roominess; stakes, poles, and guys aren't always needed if tent is to be used in mild conditions, thus cutting down on weight and cost; material is held to frame by brass hooks and shock cords; ventilation is provided by large zippered door and rear window. Color: desert tan.

Weight

One-man: 11 lb.; two-man: 14 lb.; three-man: 26 lb.; four-man: 35 lb.; six-man: 44 lb.

Price

One-man: $86.50 for FR fabric; two-man: $99.50 for NFR and $103.50 for FR; three-man: $162.50 for NFR, $171.50 for FR; four-man: $217.50 for NFR, $232.50 for FR; six-man: $245.95 for NFR, $262.95 for FR. Flies range from $22.95 to $74.00, depending on size and whether they are NFR or FR.

Ice Fishing Tent (Eureka)

ICE FISHING TENT

Size

A unique two-man tent to be used directly on the ice for fishing. Floor size is 7'1" × 7'1", with a ridge height of 6'1" in the center and 5'2" at the eaves.

Frame

Aluminum, modified umbrella frame; to secure it to the ice, ½" aluminum rods with attached ⅛" rope are standard equipment. Frame is shock-corded for ease of erection.

Fabric

Heavy-duty 4-oz. FR nylon for the walls and floor.

Features

Ventilation is provided by a tunnel vent above the door; a jigging hole (cookhole) with flap cover is built right in for cutting and fishing through the ice while remaining in the tent; material is hung from frame with brass hooks and shock cords; equipment can be hung from interior hooks and rings, and a lantern hook is also provided. Color: brown.

Weight

Sixteen lb.

Price

$110

Mountain Ark (Gerry)

Gerry—An Outdoor Sports Company
5450 North Valley Highway
Denver, Colorado 80216
(303) 292-4190

MOUNTAIN ARK

Size

Two-man: 7'4" × 4'9", with ridge height of 3'10".

Frame

Spring-loaded (with ridge pole) double A-frame, all aluminum; free-standing, although pullouts must be staked.

Fabric

Canopy: 1.9-oz. breathable ripstop; floor: 2.7-oz. waterproof taffeta; fly: 2.5-oz. waterproof taffeta.

Features

A lightweight backpacker tent with a frame that is rigid when erected but comes apart to form a pack of only 15" × 6"; front and rear ventilation with a wide front door and rear ridge vent; four side pullouts for even more room inside; double-wall construction. Colors: floor—brown; canopy—sunset orange; fly—khaki.

Weight

Six lb., 12 oz.

Price

Available upon request.

Windjammer (Gerry)

WINDJAMMER

Size

Two- or three-man: 7′ × 8′, with a ridge height of 4′ at the center.

Frame

Tripodal exoframe (modified umbrella) made of aluminum; self-supporting, although fly should be staked; excellent wind tent.

Fabric

Canopy: 1.9-oz. nylon ripstop; floor: 2.7-oz. waterproof ripstop; fly: 2.5-oz. waterproof ripstop.

Features

High degree of wind stability; dome-like shape means more interior roominess; skylight window. Colors: canopy and fly—khaki and gold; floor—brown.

Weight

Seven lb., 9 oz.

Price

Available upon request.

Year-Round II (Gerry)

YEAR-ROUND II

Size
Two-man: 7'8" × 47", with a front height of 3'7" and a rear height of 3'.

Frame
Aluminum A-frame with cross ridge bar in front, I-pole in rear.

Fabric
Canopy: 1.9-oz. ripstop; floor: 2.7-oz. waterproof taffeta; fly: 2.5-oz. waterproof ripstop.

Features
Extended fly creates a semivestibule; ventilation is provided by large front door and large rear window; packs into 15" × 6" bundle. Colors: blue floor and gold canopy with a blue fly, or brown floor and gold canopy with a khaki fly.

Weight
Six lb., 14 oz.

Price
Available upon request.

Camponaire II (Gerry)

CAMPONAIRE II

Size

Three-man: 7' × 6'6", with a front height of 3'6", a center height of 5'5", and a rear height of 3'6".

Frame

Aluminum center A-frame with cross ridge bar and two exterior I-poles.

Fabric

Canopy: 1.9-oz. ripstop; floor: 2.7-oz. waterproof taffeta; fly: 2.5-oz. waterproof ripstop.

Features

Twelve-inch vertical sidewalls; two inside pockets; clothesline loops. A lightweight year-round tent for both mild and more demanding weather conditions. Colors: blue floor with a gold canopy and blue fly, or brown floor with a gold canopy and khaki fly.

Weight

Nine lb., 8 oz.

Price

Available upon request.

Baker Tent (L.L. Bean)

L. L. Bean, Inc.
Freeport, Maine 04033

BAKER TENT

Size

A modern version of a classic design with the same basic measurements the Baker had generations ago, with a 7'4" width, a 7'4" depth, and a 6' front height, tapering down to a 20" rear height. The attached awning stretches out to cover a 6' × 7'4" area.

Frame

Aluminum I-poles.

Fabric

Here is where the Baker of today is truly different, as 1.9-oz. coated taffeta nylon has replaced the heavy hemp canvas.

Features

Roominess; sewn-in floor in rear; ventilation is provided by the huge front door and a rectangular window in the rear; entrance covered by zippered nylon mesh door. Optional equipment includes nylon mesh for awning, adding even more room, and a nylon weather awning curtain for rain, windy days and nights. Color: green.

Weight

Eleven lb., 12 oz.

Prices

Tent: $90; screen (6' × 20' nylon mesh): $21; nylon weather curtain (6' × 10'): $12.25.

Moss Sundance (L.L. Bean)

MOSS SUNDANCE

Size

A unique, aerodynamically designed tent by Bill Moss (see next segment). Two-man; 5'2" wide at the front, 3'7" at rear, and 6'8" long. Height is 53" in the front and 34" in the rear.

Frame

Shock-corded aluminum frame that combines an A-frame with a ridge pole under tension, keeping the material tight at all times. Self-standing, requiring no guy lines, but staking can be done for true tent stability.

Fabric

Rain fly, of 1.5-oz. ripstop, is not separate but actually part of the tent; floor is 2.2-oz. coated taffeta.

Features

Good wind- and rain-shedding qualities; inside pocket for storage of small items; almost self-erecting; ventilation is provided by large front door and rear window, both of which are protected by the rain fly overhang. Colors: floor—mustard; canopy—light yellow.

Weight

Seven lb., 2 oz.

Price

$175

Eave Tents (Moss Tent Works)

Moss Tent Works
Camden, Maine 04843
(207) 236-8368

EAVE TENTS

Size

Two-man: 5'9" × 7'5", tapering to a 3'7" rear width; three-man: 6'8" × 7'7", with a 4'10" rear width.

Frame

Dome; fiberglass poles that slip through pockets and grommets in floor so that entire tent can be erected in two minutes.

Fabric

Canopy and sidewalls are 1.5-oz. ripstop; floor is 2.2-oz. waterproofed taffeta; fly (which forms two zippered vestibules front and back) is a high-thread-count .75-oz. waterproof ripstop.

Features

The rain fly forms a zippered vestibule front and back that can be opened, with half of it acting as a windbreak, or closed for extra storage room; rear window is angled so that it can be left open even when facing the storm and added to the full-sized door (the flap of which tucks into a special pocket when not in use), providing excellent ventilation; sewn-in net pocket; shock cord around perimeter of tent insures tautness. Colors: floor—dark tan; canopy and sidewalls—light tan; fly—off-white.

Weight

Two-man: 6 lb.; three-man: 7 lb.

Prices

(Prices include vestibule/rain fly.) Two-man: $190 (frost liner: $27.50; snow flaps: $25); three-man: $220 (frost liner: $30; snow flaps: $27.50).

Trillium (Moss Tent Works)

TRILLIUM

Size

A six-man tent that weighs just over 13 lb. and has three vestibule-like sections that join in at the center. Each vestibule is 5'7" × 3'3", with a door height of 42". Center height is 65".

Frame

Each vestibule is a modified dome, constructed out of fiber-glass and aluminum poles connected by ferrules. The entire tent is self-standing, with even the fly attaching to the legs instead of having to be staked to the ground.

Fabric

Canopy: 1.5-oz. ripstop; floor: 2.2-oz. waterproof taffeta; fly (which fits over entire structure): .75-oz. waterproof ripstop.

Features

Excellent roominess for the weight; each vestibule has its own "No-See-Um" mesh-covered entrance with storm door, and these, added to the overhead ridge vent, provide a four-way ventilation in any weather; cookhole and overhead exhaust vent; can be erected by one person within six minutes; aerodynamic design means a free-standing tent for all weather and all environments and terrains. Color: floor—dark tan; canopy and walls—light tan; fly—off-white.

Weight

Thirteen lb., 6 oz.

Price

Including fly: $385; frost liner: $48.50; snow flaps: $42.50.

Optimum 200 (Moss Tent Works)

OPTIMUM 200

Size

A tent that is also a summer guest house that measures 7'6" high at its center and 6'5" high at the arch of each interior end of the three alcoves. Each alcove is 7'10" long, 8'6" wide at the center, and 5' wide at the end. The center of the tent is 7'6" from alcove to alcove.

Frame

Pre-bent, curved aluminum with aluminum center I-pole for added stability.

Fabric

Nine-oz. cotton duck, with a waterproof nylon floor.

Features

A large, walk-through door with awning in one alcove and large windows in the other two provides constant and excellent ventilation; decorative cotton lining (price available upon request) can be attached to interior sidewalls, making the Optimum 200 an all-weather, all-season tent; can be used with or without platform; floor can be carpeted.

Weight

Forty-five lb.

Price

$960

Optimum 200 (interior, showing carpeted floor)
(Moss Tent Works)

Optimum 200 (interior, showing optional liner) (Moss Tent Works)

Van Tent (White Stag)

White Stag
Hirsch Weis
Division of Warnaco, Inc.
5203 S.E. Johnson Creek Boulevard
Portland, Oregon 97206
(503) 775-4314

VAN TENT

Size
Used either as an extension of a side-door van or as a separate tent, it measures 7'10" × 6'9" and will sleep three to four. Ridge height: 6'6"; wall height: 6' even.

Frame
Adjustable aluminum box.

Fabric
Canopy: 6-oz. canvas; sidewalls: waterproof synthetic duck; floor: heavy-duty waterproof nylon.

Features
Front and rear full-sized doors, with the back entrance attaching to the open sliding doors or even to the swing-out type of doors of a van with magnetic tape; three windows, two on the side and one in the front door, for complete ventilation; with proper guying and staking, tent can be used by itself as a family shelter; can be left standing while van drives away. Color: brick red.

Weight
Twenty-seven lb.

Price
$223

Van Pickup Tent (White Stag)

VAN PICKUP TENT

Size

For use with vans or pickup trucks with cab-high canopies, it measures 7'11" × 6'10", with a 6'8" ridge height and a 6'2" wall height.

Frame

Aluminum box.

Fabric

Canopy: 6-oz. canvas; walls: waterproof nylon; floor: double-coated, waterproof ripstop.

Features

Two full-size doors, front and rear, with the back entrance attaching to the side of a van or the rear of a cab-high canopy of a pickup truck with magnetic tape; three windows, two on the sides plus one in front door, for ventilation; tent can be left standing while vehicle drives away and can be erected without the truck or van. Color: green.

Weight

Twenty-five lb.

Price

$158

High Lake IV (White Stag)

HIGH LAKE IV

Size

A small, backpacker version of the roomy sidewall tent. Although light in weight, it measures a full 7' × 7' with a center ridge height of 5' and a ridge length of 7'. Sleeps 3 or 4.

Frame

Aluminum front and rear I-poles.

Fabric

Canopy and sidewalls: waterproof ripstop nylon; floor: heavy-duty, waterproof ripstop sewn right to sidewalls (unlike the original sidewall, which didn't have a floor).

Features

Large front doors are covered with fine nylon mesh; has storm flaps that can be tied over the entrance to keep out the wind and rain or lashed back to catch gentle breezes; rear window offers good cross ventilation; 6 pullouts, 3 on each side, for even more inside roominess.

Weight

Six lb., 12 oz.

Price

$52

It should be repeated once again that the tents featured here are just selected models and do not represent the manufacturers' complete lines. Most tentmakers market shelters that run the gamut from the huge family size tents to tiny, one-man shelters that fit in a pocket. Featuring *all* the tents on the market would be repetitious and would require a book ten times the size of this one. Therefore, it would be wise to write to any or all of the tent designers and ask for their catalogs. Using them and *The Complete Tent Book* will provide all the information you need to satisfy your tent needs and wants before you leave the comfort of your living room.

The following is a listing of manufacturers who answered too late to have their tents included in the preceding section:

Pacific/Acente
P.O. Box 2028
2126 Inyo St.
Fresno, CA 93718

Eddie Bauer
Third & Virginia
P.O. Box 3700
Seattle, WA 98124
1-800-426-8020 (toll-free number)

Recreational Equipment, Inc.
P.O. Box C-88125
Seattle, WA 98188
1-800-426-4840 (except Washington, Alaska,
    Hawaii)
1-800-562-4894 (for Washington residents only)
1-800-426-4770 (for Alaska and Hawaii)

8

Other Shelters

There may be times when the weather forecasters assured you that the skies will be clear and sunny for days and one glance up at the total absence of clouds gives you confidence in those predictions—until the heavens suddenly split apart right in the middle of your Sunday hike and the heaviest rain in eight generations drenches everything, including you.

There may be occasions when the sun beats down on the landscape day after day and you may want a little relief from its burning rays for your on-the-trail picnic, some shelter that would provide plenty of clean, cool shadows as well as be open enough to capture the slightest breeze.

Or you may be one of the rare individuals whom mosquitoes and other bugs avoid like DDT and you don't think you need a tent with a tub floor and nylon mesh to protect you from woodland insects.

In any event, there may be periods when you don't need or want a full-sized tent and will settle for some kind of emergency protection just in case. This chapter will detail what's available in this category of other shelters.

Of course, you could just carry one of those instant one-man plastic tents that fit in a pocket or glove compartment. But to use that type of shelter with a family does mean a few problems. You'd have to carry one for every member of the group, and the kids aren't always handy when it comes to laying out even a simple instant tent. Or you could utilize the tube tent, but this shelter lacks proper ventilation and is often as uncomfortable inside as it is on the outside in both rain and sun. Or you could carry along a discount store plastic sheet and a few versa-ties, but these are hard to rig and aren't built to withstand the strain a shelter would have to take in an average rainstorm. Consequently, if you're interested in other shelters, you should consider a waterproof plastic

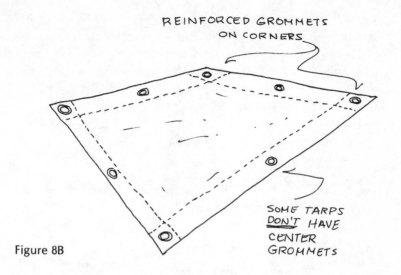

REINFORCED GROMMETS
ON CORNERS

SOME TARPS
DON'T HAVE
CENTER
GROMMETS

Figure 8B

tarp specifically designed for that purpose. The key word is *designed;* even tarps have had creative minds revamping them.

For instance, there's a style of tarp that has reinforced grommets at the four corners and in the centers of each side (see Figure 8B). These holes allow the tarp to be rigged in a variety of ways so that it will afford adequate protection in just about any adverse weather condition (see Figure 8C).

Another example is a tarp marketed by Eureka! that features ties attached at various points of the tarp (see Figure 8D). These reinforced strips of canvas function as connections for guy lines, stakes, rope or tree frameworks, or almost anything, making this tarp one of the most versatile designs available.

The sizes of these tarps vary, but usually both the grommeted model and the tie-down are about 10' × 10', weigh in the neighborhood of 2 to 2½ pounds, are made of (again, *usually*) 1.9-oz. waterproof nylon, and cost from $20 to $35.

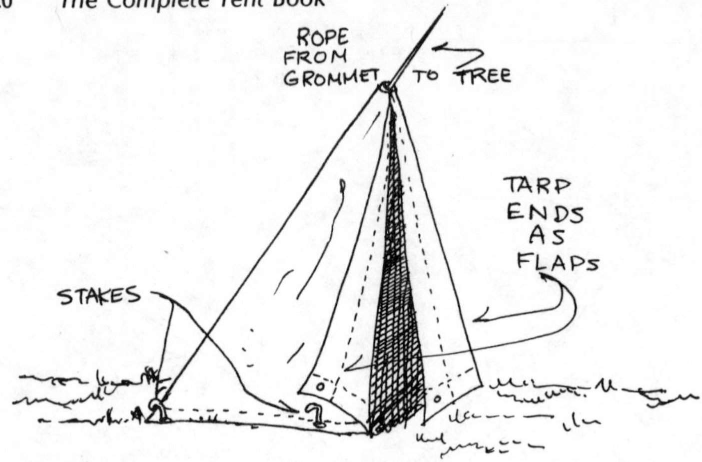

ROPE FROM GROMMET TO TREE

TARP ENDS AS FLAPS

STAKES

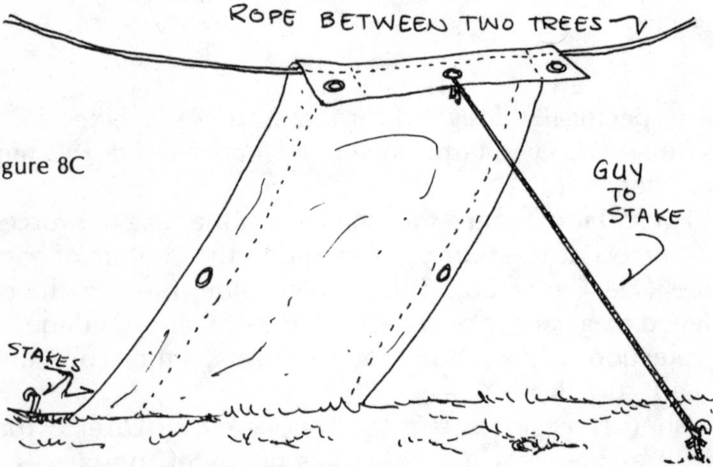

ROPE BETWEEN TWO TREES

Figure 8C

GUY TO STAKE

STAKES

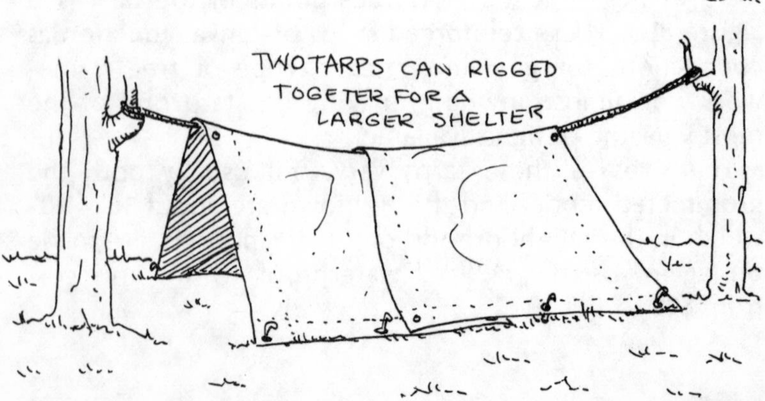

TWO TARPS CAN RIGGED TOGETER FOR A LARGER SHELTER

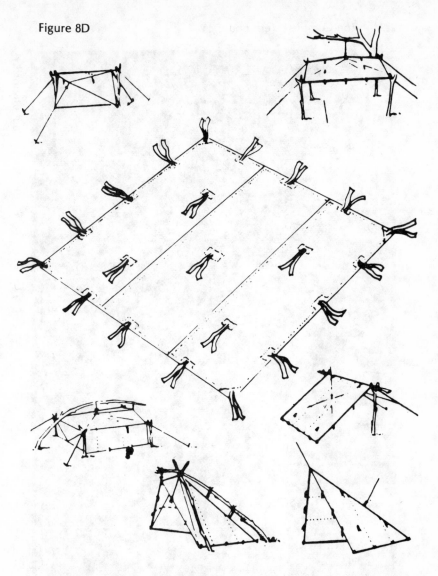

Another illustration of how an inventive mind can revamp what started out as a single hunk of plastic is the Parawing by Bill Moss of the Moss Tent Works in Camden, Maine (see Figure 8E).

Figure 8E

(Photo by Paul Oratofsky)

The Parawing is a hyperbolic paraboliod, which is really a plane surface that has two high points and two low points. The Parawing's shape facilitates fast water runoff and spills wind effectively. The Parawing comes in a 12-foot model, as measured from high point to high point, made out of waterproof nylon; a 19-foot model, made out of the same fabric; and a 19-foot model of Kimberly Clark's new Evolution II material, a synthetic fabric of spun-bound polypropylene fibers (polypropylene is the plastic used in containers and electrical connections). The prices are: $30 for the 12' (poles extra), $72 for the 19' nylon (poles included), and $40 for the 19' Evolution II (poles included).

Another piece of equipment a smart hiker should carry and that can pull double duty as an emergency shelter is the poncho (see Figure 8F). This versatile bit of raingear can be used singly, or two can be snapped together to form a larger shelter; both single or double ponchos can be rigged like the tarps. It weighs only 12 or 13 ounces on the average and costs from $10.95 to $30, depending on brand name, styling, and other features. But even the inexpensive ponchos are worth the price in an emergency.

There are other shelters that could be utilized in an emergency—do-it-yourself lean-tos made out of tree limbs, four-posted shelters constructed out of stripped trees and roofed with cut limbs, and other structures that utilize natural foliage. These shelters are frowned upon— unless it's an absolute life-or-death situation—by all experienced campers and park and wildlife officials. Because tarps and other emergency shelters are so compact and lightweight, there's no reason for anyone, no matter how inexperienced in the woods, to be caught without some kind of bring-along, easy-to-carry, inexpensive substitute shelter. No reason at all.

Figure 8F

9

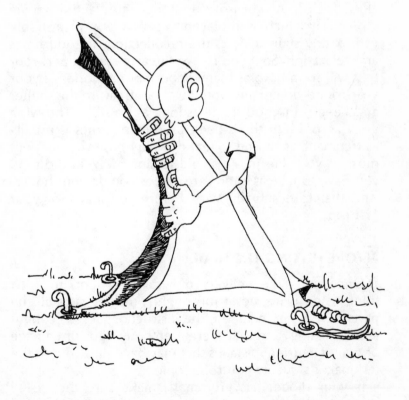

Loving Your Tent

Naturally, the most enjoyable part of owning a tent is using it. But beware: a strange thing often happens to campers if they spend enough time in their structures. They tend to first like and then love their tents, and they frequently start getting the same warm tingle about their temporary soft homes that they get when thinking about their permanent hardwood houses. Most campers eventually look at their tents with fond memories. A tiny rip that is bound to happen is associated with the happy moments of a child growing up or a silly mistake of a close friend. A worn-out spot becomes a tiny movie screen on which is replayed a particularly pleasurable trip. A tiny stain triggers the recollection of good times and relaxation. So if you're like the average camper, your tent will soon have as many important memories as your year-round home, and you'll want to treasure the shelter and keep it in good shape as long as possible. Throwing away an old tent that's been used many times is usually as traumatic as having to tear down a house you've spent most of your life in. To avoid tossing away an old tent, you have to take as good care of it as you do your house. And that care actually starts before you leave for your first trip.

BEFORE HITTING THE TRAIL

Once you take delivery of your new tent, the first thing to do is inspect it thoroughly, making sure all the parts have been included, the instructions for erecting it aren't missing, and that there aren't any signs of damage or manufacturing mistakes that may have slipped past the tentmaker's quality control people.

If your model has grommets, make sure they're all there and that they are centered in the reinforced patches, tightly connected, and can't be turned in the holes. Determine that the other reinforcements, whether they're patches of webbing, cloth sewn to the fabric, or just extra stitching, are where they are supposed to be and are tight, with no loose ends. Slide the poles together to see if they jam or if the snap-locks stick in the slots. If jamming is a problem, check for burrs; if you find any, usually a little filing will eliminate them. It would be a good idea to apply a *very light* coating of oil, as this should do away with pole- or lock-jamming. Count the poles and accessories to be sure they're all there, and go over the directions for erecting the shelter. Then take it out in the backyard and put it up, remembering to follow the instructions without any ad-libbing.

Once the shelter is up, check the line. If you spot too many wrinkles (a few are natural in a brand-new tent; they disappear after the fabric has become "conditioned"), check the guy line. Often, too much tension causes as many wrinkles as not enough. Keep playing with the guys and stakes—if they're used with your tent—until the structure is as smooth as you can get it. Then take it down and do it all over again as soon as you can. This serves two purposes: it helps you become familiar with the proper procedure for erecting your particular model, and it helps you learn which is the best way to guy and stake it.

With canvas tents, there's one more step before you take down the structure and haul it to your site. As explained by Eureka!: "A new tent may leak slightly until the canvas is 'conditioned.' It should be hosed down and dried twice before using in order to shrink the fabric and set the shape of the tent."

Actually, it would be good to hose down a nylon tent too and then crawl inside to see if the seams leak anywhere. If you find a drippy spot, cover it with wax or the seam sealant that usually comes with the tent. In fact, sealing all the seams would be a good idea. Let the tent dry completely before you pack it away.

ON THE TRAIL

Unless you're heading for established campgrounds with running water and electricity, you should know something about the area you're going into. This is where the topographical maps mentioned earlier are so valuable. They truly give you the lay of the land. Other ways to choose a site are to check into local regulations regarding pitching tents near hiking trails and running water (recently, the National Park Service instituted a regulation prohibiting any tent pitching within one hundred feet of the shore of a stream, river, or lake to protect the more delicate flora that grows in these spots), campfires, and other rules applying to camping.

It's also smart to avoid the bottoms of ravines and gullies, since they can become flooded from storms that occur miles away. Also, because cold air settles, they are the depositories of chilly air during the night. A point to remember is that halfway up a hill is better than the top (where the winds are stronger) or the bottom (where it's colder). Use natural windbreaks like trees, brush, and other landscaping. Experience will guide you in further site-choosings.

Finally, when you've selected a site, there are three basic steps to pitching the tent. Gerry Cunningham recommends, "First, lie down on the area selected and move around until the ground feels comfortable. Pitch the tent over this spot.

"Stake out the floor, very tightly and as square as possible. If the floor has diagonal wrinkles, it is not perfectly square.

"Erect the head end with poles and guy lines (if your tent is the "A"-shaped, backpacker type; otherwise, see your instructions). The fabric will then drape from the peak to each of the rear corners. If one side sags more than the other, something is wrong. Maybe the guy line is pulling to one side or the other even though it looks straight. Maybe one corner is lower than the others. Before you go on, get the tent to drape symmetrically in this half erected position or it will never be flat later.

"Erect the other end after the head end is to your satisfaction. Any wrinkles that show can usually be corrected by the angle and the tension of the guy line. The other guy lines are now pulled out, symmetrically, a pair at a time."

If you'd like this booklet that goes into many details bout backpacking, you can write to Gerry and ask for *How to Enjoy Back-Packing* by Gerry Cunningham.

Other on-the-trail tips include adjusting the net if it's going to be in one place for more than one day; sweeping it out with a whisk broom constantly, since twigs, pebbles and other forest debris brought in by the users can cause abrasion wear-spots on the floor; taking off hiking boots before entering the shelter; and immediately mending any tears or worn spots you find with a repair patch kit, fabric cement, or repair tape, items that should be part of your equipment whether you are backpacking or camping in a huge, van-size tent.

The sun should be kept from beating on the tent, especially if it's a nylon one, either by pitching it in the shade or by using a rain fly. The tent should be dry when it's repacked, especially with canvas models. If you should have to roll up a damp tent for some reason, try to dry it out as soon as possible. Also, sweep out the inside *and* the outside as you roll it up. Twigs caught in the strands can trigger rips or puncture holes in the toughest of fabrics.

AFTER THE TRAIL

Once you're finished with your shelter for the time being or at the end of your particular camping season, you should wash off the tent by first either rigging it in

your backyard or even in the shower. Some manufacturers claim you can wash the tent in machines, but if you'd
rather be safe and wash it the old-fashioned way, either
pitch it and hose it down after sponging it with warm
water and a gentle detergent or rig it from the suspension points in the bathtub and, again, sponge it down
with warm water and a soft detergent. Again, the instructions that come with the tent usually recommend cleaning methods and soaps to be used.

Let the tent dry thoroughly and then check it for rips
and worn spots. Repair any you find. Once it's dry, clean
and lightly oil the poles. Then, as Eureka! suggests, "All
tents should be stored dry, folded loosely, in a cool, dry
place. Keep it out of reach of mice, as they like to nest in
tents."

And how long should your tent last?

Again, Eureka!: "Average campers use their tents 21
days per year. With this kind of usage, it will last many
years. Ten- to fifteen-year-old tents are not unusual.

"A tent erected for the whole season may only last one
to three years, as wind, rain and sun will have detrimen-
tal effects on canvas and nylon."

In other words, love your tent and it will love you.

Index